CW00555564

Natural Ingredients in Cosmetics

REVIEW COPY

Natural Ingredients in Cosmetics

Based on papers presented at a symposium entitled "Natural Ingredients — Fact or Fiction?", which was organized by the Society of Cosmetic Scientists and held at the Park Lane Hotel, London on May 23, 1989

edited by

Marian Grievson, Janet Barber and Anthony L.L. Hunting

 **MICELLE PRESS**

Weymouth, Dorset, England and Cranford, New Jersey, USA

Copyright © 1989 and 1993 by the Society of Cosmetic
Scientists

All rights reserved. Apart from any fair dealing for the purposes of
research or private study, or criticism or review, as permitted
under the UK Copyright Designs and Patents Act, 1988, no part of
this publication may be reproduced, stored in a retrieval system,
or transmitted, in any form or by any means, electronic, mechani-
cal, photocopying, recording or otherwise, without the prior permission in writing of the publishers, or in the case of reprographic
reproduction only in accordance with the terms of the licences issued by the Copyright Licensing Agency in the UK, or in accordance with the terms of licences issued by the appropriate
reproduction rights organization outside the UK.

A catalogue record for this book is available from the British
Library

Published simultaneously in the United States of America
and Great Britain by

Micelle Press, Inc.
P.O. Box 653, Cranford, NJ 07016, USA

Micelle Press
12 Ullswater Crescent, Weymouth, Dorset DT3 5HE
England

US Edition: ISBN 0-9608752-6-3
UK Edition: ISBN 1-870228-07-3

The information given in this book is presented in good faith, but
no warranty, express or implied, is given nor is freedom from any
patent to be inferred

Printed and bound in Great Britain by Antony Rowe Limited
Chippenham, Wiltshire

Editors' Note

This book is an edited version of some of the papers read at a symposium held at the Park Lane Hotel, London on May 23, 1989, and which was entitled "Natural Ingredients — Fact or Fiction?" The symposium was organized by Marian Grievson on behalf of the Society of Cosmetic Scientists and we are grateful to the Society for permission to reprint these papers here.

Acknowledgements

The drawings used on the cover and on pages 59 and 61 are by Patricia Margret. Other illustrations are either by the chapter author(s) or supplied by their respective companies.

Contributors

M. Benhaim, Jean Bommelaer and **B. Levrat**

Exsymol S.A.M.
4 avenue Prince Héréditaire Albert
MC 98000
Monaco
(represented in the U.K. by Paroxite [London] Limited, Office Unit 2, 7 Dryden Court, Renfrew Road, London SE11 4NH)

Janet M. Barber

Micelle Press
12 Ullswater Crescent
Weymouth
Dorset DT3 5HE

Martyn Davies

ASDA Stores Limited
Great Wilson Street
Leeds LS11 5AD

Anthony C. Dweck

Peter Black Cosmetics Limited
Frogmore House
Frogmore Road
Westbury
Wilts BA13 4AT

Marian M. Grievson

Créations Aromatiques
Bishops Palace House
Riverside Walk
Kingston On Thames
Surrey KT1 1QN

Bryan Grimshaw

Bush Boake Allen Limited
Blackhorse Lane
London E17 5QP

Dr Keith Helliwell

William Ransom & Son plc
104 Bancroft
Hitchin
Herts SG5 1LY

Anthony L.L. Hunting

Micelle Press
12 Ullswater Crescent
Weymouth
Dorset DT3 5HE

Malcolm James and **David Mitchell**

Chesham Chemicals Limited
Cunningham House
Westfield Lane
Kenton
Middx HA3 9ED

Peter Jarvis

Peter Jarvis Cosmetic Developments Limited
8 Hadleigh Business Park
Pound Hall Road
Hadleigh
Suffolk IP7 5PW

Roy E. Nixon

D.F. Anstead Limited
Radford Way
Billericay
Essex CM12 0DE

Society of Cosmetic Scientists

Delaport House
57 Guildford Street
Luton
Beds LU1 2NL

Contents

Introduction*

Martyn Davies

How many of you buy your own toiletries and cosmetics? How many of you buy only those containing "natural" ingredients?

My own answers would be, Yes, I do, to the first question and, No, I don't, to the second. As with many consumers, performance and price affect my decisions. "Natural" is a bonus. Does the consumer really care, then, or are natural ingredients another marketing gimmick dreamt up to bolster sales, increase profits, keep shareholders happy and ensure jobs for cosmetic scientists?

Certainly there are people who would answer in the same way and no argument, rational or otherwise, is likely to change their minds. These are a minority. The majority are prepared to be persuaded.

This is clearly demonstrated by the growing concern for the environment. Movements like Green Peace and Friends of the Earth have gained increasing support for their views and consumer groups also have accelerated their activities. They all used to be thought of as extremists, but that attitude has now altered: the rapid change from the use of chlorofluorocarbons in aerosol products is an example. At one time the industry would have withstood such pressure — not any more.

The media have had and continue to have a significant influence on the public. Recently, media attention to issues such as salmonella in eggs and poultry, listeria in soft cheeses and pesticides, like daminozide (Alar) on apples, has provoked public

---

*Based on a speech introducing the Symposium and originally entitled "Natural Ingredients — The Dilemma".

concern. Dioxins have also caused difficulties in the non-food industries. There has been, in each case, some element of fact which has been exaggerated, causing alarm and worry to consumers initially, but culminating in their deciding to carry on as before since whatever they do or eat will kill them — an attitude satirically summarized by Private Eye in an article headed "Man Eats Meal and Lives, Shock".

Unfortunately, alarmist reporting has resulted in much cynicism, so that when a truly major problem occurs, the public is led to underestimate its seriousness. However, if you talk to any company involved in scares, you will probably be told, though perhaps only in private, that sales went down. The power of the media should, therefore, never be underestimated. Bad news still sells newspapers because we, the public, like bad news.

Yet consumers clearly want to be better informed, as was evidenced by the number of visitors to the ASDA Festival of Food and Farming recently* held in London. On the Saturday nearly 3,000 people per hour visited the ADSA stand to be better informed. They were astounded to find, for example, that despite advertising implying that biodegrabable detergents were a new invention, detergents had been biodegradable since the 1960s. An interest in the question of natural or non-artificial products was much in evidence, and was even more pronounced with regard to food. Schoolchildren were the most knowledgeable, certainly more so than their parents, who in turn were more so than their grandparents. Schoolchildren were also more demanding in their insistence on detailed replies, an attitude echoed in the enquiries made at our stores and received directly at our Head Office. I am sure your own experience is similar. Many consumers are suspicious, and therefore plastering all products with the label "natural" would certainly cause a stir. Yet many companies, manufacturers, raw materials' suppliers and retailers successfully deal with predominantly natural products. What then, is the problem? The need for a symposium and the number of delegates present suggest that there is a problem.

---

*This introduction was originally presented in 1989. — Eds.

What does *natural* mean? My dictionary defines it as: "pertaining to, produced by or according to nature; furnished by or based on nature; not miraculous" — well, so far so good. It then goes on: "not the work of man, not interfered with by man". Ah! let us move swiftly on.....

A definition of natural resources might be more appropriate to our purposes. I suppose a simple definition might be "minerals, plants and animals"! I had better not dwell on animals. Consumers do not like animal testing and are getting less tolerant of animal ingredients. Will tallow in soap become an ingredient of the past? Even beeswax and lanolin are seen as representing exploitation of insects and animals.

Minerals and plants are therefore all that is left to use. There is certainly no dearth of these, if we exclude coal and oil, resources which, we are told, are running out and being replaced by nuclear fuel.

How are we to utilize our natural resources? Can plants be grown specifically to meet our needs? Is it natural to encourage growth and good yield with fertilizers and pesticides? Can the crop be harvested other than by hand? How can the natural ingredients be extracted from the natural resource? Is pressing the only way? Is steam distillation allowed, and what about solvent extraction? If that is allowed, what about other chemical means?

These questions may appear to be irrelevant to you and me, but to consumers they are basic to their understanding of what is natural. Then come the questions about the finished product.

How much of the product is natural? How much of any natural ingredient is present? You put in only enough to make a marketing claim and, as you can detect at parts per trillion, there can't be a benefit, can there? Is it safe? Are the products effective? How do you know? To answer "by animal testing" to that question can be commercial suicide for "naturals"! Consumers who know a little more then start to get difficult. These "naturals" come from third world countries, so aren't you using cheap labour and then charging high prices? A lot of these naturals use up natural resources, so aren't you effecting an ecological imbalance? The latter question is commonly asked about pulp products and consumers are quite often heartened to hear that more trees are planted than cut down.

The public has learnt a lot about natural products, particularly with respect to food. More and more food products state that they contain no artificial flavourings, colourings or preservatives. I was, however, recently surprised to discover that a friend believed that no artificial flavourings meant that the product was totally natural, and to witness her disappointment on finding this to be untrue.

Food is governed by three principal pieces of legislation: the Food Act, 1984, the Food Labelling Regulations, 1984 and the Trade Descriptions Act, 1968. The last-named obviously covers all products, both food and non-food.

In talking to my food colleagues, I found that no legislation exists at present that specifically governs natural ingredients. A document was circulated by the Ministry of Agriculture, Fisheries and Food including proposals from the Food Advisory Committee for discussion in December 1987. These discussions are still taking place as agreement is difficult to achieve. We did not agree with all the proposals. For example, under the proposals, "natural" yoghurt would no longer be permitted, although the term has been used to describe plain yoghurt for years.

The proposals were aimed at reducing the confusion customers experienced in the area of food labelling, particularly as there are practical difficulties in enforcing the Food Act, because interpretation of a label is a matter of subjective judgment. It was hoped that voluntary self-regulation would occur, but this has not yet happened.

In the United States the Food and Drug Administration (FDA) has taken no position in its use of the term "natural" on food labels as natural ingredients are so difficult to define.

I doubt, however, that consumers will be satisfied with the absence of any specific regulations. While they may be able to console themselves with at least knowing what ingredients are in processed foods and can estimate their amount by their position in the ingredients list, they cannot do so with regard to cosmetics and toiletries, at least not in the United Kingdom. The labelling of ingredients in cosmetics and toiletries is, I believe, likely to become obligatory, especially in view of the introduction in 1992 of a single European market. Judging by the number of requests we receive for ingredient information for our products, the consumer lobby is likely to become more vociferous on the matter.

The consumer, meanwhile, is left confused. Some of you may believe that the consumer has enough protection and information now that we all have to conform to the Trades Descriptions Act, 1968, the Cosmetics' Directive and the Consumer Protection Act, 1987. Trading standards officers feel justified in accusing certain members of the food industry of telling half-truths and untruths with respect to "naturals". It is to be hoped that they do not put any cosmetic products under the same umbrella. But is hope enough? Can we do anything to prevent such accusations? Do we have to live with the nagging doubt that when we pick up a newspaper or turn on the television or radio we might learn of a "natural" scandal? It might not be true, but try telling the media that after the event.

What then does the consumer want? To be informed — not misinformed.

Any information should be consumer-orientated and meaningful, helpful at the point of sale. Consumers do not want their fears encouraged or exploited — they do not want to be confused. It is easy to sell a product once, but the customer is not stupid, he or she does not like being conned — I certainly don't!

Without customers, retailers like my company would not exist, nor would manufacturers or raw materials' suppliers.

We therefore have a choice: to be reactive to crises or to be proactive. Should we, for example, consider a voluntary code of practice for labelling ingredients before it is forced upon us?

I have asked many questions and provided perhaps few answers — for that I make no apologies, as I have tried to reflect some of the concerns of all our consumers. They are in a difficult position. I hope that some of the questions I have raised are answered by the papers that follow.

Natural and Nature-Identical Fragrances — A New Creative Challenge

Bryan Grimshaw

Synopsis

A demand for a fragrance created from materials that are entirely "natural" or even "nature-identical" places considerable limitations on the perfumer, setting his creative freedom back a hundred years. It can also have a dramatic effect on costs. Such fragrances cannot even be said to be safer, so are they worthwhile?

Fragrances are different from many of the other ingredients in use in cosmetics and toiletries today because they are themselves mixtures of many materials, as the following fragrance formulation shows:

Jasmin Fragrance

Bergamot Oil (bergaptene-free)	20
Jasmin Absolute	1
Petitgrain Oil Bigarade	1
Ylang-Ylang Oil Extra	10
Linalol (ex Ho Oil)	8
Eugenol (ex Clove Oil)	1
Benzyl Acetate	90
Benzyl Propionate	9
Geraniol 980	20
Lilestralis	6
Hexyl Cinnamic Aldehyde	14
Abbalide	20
	200

A typical fragrance will contain from 10 to well over 100 ingredients, almost always a blend of natural and synthetic substances. The simple fragrance formulation shown above contains totally natural materials, "nature-identicals", and totally synthetic aroma chemicals. Very few fragrances in use today are totally natural and even then we must define what we mean by the word "natural".

For thousands of years, until about 150 years ago, all fragrances were wholly natural. The only materials available to perfumers were the essential oils, resins and absolutes obtained from plants, and the few animal materials with odour value. Moreover, these materials were all obtained by physical methods of separation and thus were not changed chemically during extraction. Thus a definition of a natural fragrance was unnecessary. Today, before we can discuss the definition of a natural fragrance, we shall have to look in more detail at how the individual components are derived.

The impact of chemistry

In the 1830s, chemists began to isolate and identify single aroma chemicals from the natural oils in which they occurred, for instance, pure cinnamic aldehyde from cinnamon oil and benzaldehyde from almond oil. These separation techniques are still being used today to produce "isolates", which, although only a part of the natural fragrant extract, are still considered natural.

Then, from the 1850s, aroma chemicals began to be synthesized, at first in the laboratory and later on a larger scale, by chemical reaction from different starting materials. Initially the chemists sought to duplicate known aroma chemicals, but soon new molecules with odour value were discovered that were not known to exist in nature.

Although at the beginning of the twentieth century over 90 per cent of fragrance raw materials were still natural, they had been joined by two other groups of chemicals whose use has gone on increasing ever since. These were, and are, "nature-identical" materials, which exist in nature as components of plant essential oils, etc. but which have been produced by chemical reaction rather than by physical extraction/concentration, and "synthetics", which are substances with odour value that do not occur naturally. It would be more correct to say of the latter group of materials that they are not known to occur in nature; discoveries of chemical species previously unknown in nature continue to be made by virtue of improved analytical techniques and the investigation of different plant extracts. For example, hydroxycitronellal, which is of major importance in muguet (lily-of-the-valley) odours and which was unknown until it was chemically synthesized, has now been discovered by Chinese scientists in the blossoms of three plant species. Amyl cinnamic aldehyde was also believed not to exist in nature, but again it was recently reported to have been identified in a plant extract.

Today the majority of perfumery raw materials are synthetically produced. A typical cosmetic fragrance may contain only around 25 per cent of naturals. So let us ask the question why perfumers now use fewer natural materials and more synthetics? The answer will reveal why a request for a natural perfume today is more of a challenge than it was 150 years ago.

Availability of natural materials

The demand for fragrances continues to increase. Today the world consumption of perfumes is probably in excess of 150,000 tons per annum. This increase is due in part to an increasing population, but particularly to a larger population able to afford perfume and perfumed products. Personal products, which concern us today, have almost always been perfumed anyway and we have seen the number and diversity, and the use, of these products grow dramatically.

Unfortunately, the economic growth mentioned above has led if anything to a decline in the availability of many natural plant oils which were often produced on a small scale by very labour-intensive processes; they were often little more than cottage industries.

With a few exceptions, production of these plant oils has been unable to keep pace with demand. By contrast, synthetic aroma chemicals use starting materials, often more than one, that are in plentiful supply.

The cultivation of plant species will always be at the mercy of the weather; yields can vary from year to year, and quality, which means the odour, can also fluctuate. Economics and politics can cause the abrupt abandonment of specific cultivations, particularly as many of these plants are grown in the less developed countries. As a consequence the price can fluctuate wildly but tends gradually to increase, reflecting lower availability. The few exceptions are some of the citrus and mint oils, and clary sage (there may be others), which are now produced by modern "farming" methods. Significantly, these tend to be materials with major use in flavours.

Biotechnology may change this situation, but the large-volume production of fragrance materials by this technique seems some way off. In the meantime, synthetic aroma chemicals are much more stable in terms of availability, quality and price and it is certain that the volume demand for perfumes cannot be satisfied from natural sources alone.

Novel odour types

Perfumers are artists and are always striving to create a new and different fragrance. Some synthetic chemicals have odours which are unknown in nature and therefore assist creativity. Pure fantasy notes are popular in personal fragrances, although many are of course related to naturally occurring odours.

Synthetics also allow the perfumer to replicate some types of odour for which natural extracts are not available. To return to our example of the muguet odour; no lily-of-the-valley oil is available and this beautiful odour type can only be achieved by the use of similar smelling molecules such as hydroxycitronellal.

Most "classic" perfumes could not be made without synthetic chemicals. Chanel No. 5, created in 1921, was the first alcoholic fragrance to contain important amounts of synthetic material — the aliphatic aldehydes. Since then benzyl salicylate has allowed the creation of L'Air du Temps; Eau Sauvage depends on methyl dihydrojasmonate, Cabochard on isobutyl quinoline, and so on.

All of today's alcoholic fragrances use synthetic aroma chemicals, and most cosmetics and toiletries, including those that promote a strong natural image, do not have "natural" odours and are certainly not perfumed with totally natural fragrances. There are many odour types with which we are familiar, which are impossible to produce without the use of synthetically derived aroma chemicals.

Chemical stability

Synthetic aroma chemicals can be "engineered" to be stable and perform well in aggressive bases. Even in cosmetics and toiletries, acidic, alkaline, oxidizing and other potentially reactive conditions are encountered. The perfumer must guard against his fragrance being chemically broken down by the base — at best weakening it, but at worst causing off-odours. A single pure chemical, synthetically produced, may be more predictable in behaviour than a natural oil or even an isolate, both of which may contain a large number of different chemical species, all contributing to the odour and all potentially reactive.

Odour performance

Last but not least, a perfumer can use synthetic chemicals to make fragrances more lasting, especially on the skin. Nature's fragrances are beautiful but often transient. They are continually being produced by the natural source, whereas we often want a fragrance to last for many hours on the skin following one application. It is particularly when persistence of fragrance in use or after use is required that a perfumer tends to employ synthetic fixatives, especially the man-made musks, to give tenacity.

I think you can begin to see why I said in my synopsis that to ask a perfumer for a natural fragrance places great restrictions on his creativity. The fragrant materials of animal origin — amber, musk, castoreum and civet — are all but denied him by virtue of price constraints and humanitarian principles and many of the plant species which he values are not now grown on the scale of a century ago. There are likely to be many more demands placed upon the fragrance today in terms of performance and stability. In fact a request for a natural fragrance is a much greater challenge to his creativity than most fragrance briefs.

Having identified some of the problems of producing, and using, totally natural fragrances, let us spend a few minutes on the benefits of naturals. Even the simplest fragrance formulation such as we saw earlier is likely to contain a proportion of natural materials. This is because the "naturals", being themselves mixtures of many chemical species, can confer a "roundness" on a perfume. They give a richer, fuller character to a fragrance which might otherwise appear to our extremely sensitive sense of smell as just a mixture of materials, rather than as a true blend or "accord". This is a good time to look in a little more detail at the groups of materials which perfumers use.

Natural materials
Natural materials can be divided into those few that are derived from animal sources and the much greater number that are obtained from plants.

1. Animal-derived materials
There are only four fragrance materials of animal origin that have been widely used in perfumery and these are becoming less and less common now. They are amber, or ambergris, from the sperm whale, castoreum from the beaver, musk from the musk deer, and civet from the civet cat. Civet can be obtained from specially bred animals without harming them. Ambergris, however, was normally obtained after the animal had been killed and certainly it is necessary to kill the animals to obtain castoreum or natural musk. Only civet with its sour animalic odour and hint of sex appeal, and castoreum, warm, leathery and sweet, are commercially available today, and both are very expensive.

Perhaps I should qualify my statement on musk by saying that experiments are under way to obtain the material from live musk deer without having to sacrifice the animals, but there is unlikely to be a useful quantity of natural musk available for perfumery use from this source in the foreseeable future.

2. Plant-derived materials

There are thousands of species of plants that yield fragrant material (essential oils etc.), often from the blossom, but also from other parts of the plant such as its leaves, stem, root, wood, fruit, and so on. Oils and other extracts from around 200 plant species are available commercially. Some extracts are available in abundance, for instance the citrus oils such as orange, lemon, lime, etc. Others are extremely scarce, perhaps because of the cultivation difficulties but often because the yield of oil is so low. For instance, one kilo of rose oil is obtained from around five tons of rose blossoms; tuberose is literally worth its weight in gold.

One thinks immediately of blossom in relation to natural perfumes, and certainly many essential oils are obtained from the flowers. The most important are probably rose, jasmin and ylang-ylang. Neroli oil is interesting in that it is one of three oils obtained from different parts of the same plant, the bitter orange tree, in this case from the flowers as they are just about to open. Neroli is bitter-sweet and spicy. Ylang-ylang has a very diffusive sweet floral odour, reminiscent of jasmin. A rose is a rose, but often with green top notes and tea or honey-like nuances. And jasmin — a lot nicer than benzyl acetate, its main constituent.

Occasionally an oil is obtained before the flower opens, for instance clove bud oil is obtained, as the name suggests, from the dried flower buds. Again another oil, clove leaf oil, can be obtained from the leaves and twigs of the same plant. Clove bud oil is warm, spicy and sweet, whereas the leaf oil is more "dry".

Also from this source, i.e. leaves and twigs, come such oils as eucalyptus, geranium, patchouli with its intense woody and balsamic odour, and petitgrain, the second oil from the bitter orange tree. There are several species of eucalyptus from different parts of the world yielding oils of varied character, not only camphoraceous but with rosy, grassy, minty and lemon-like notes. Similarly geranium can come from Morocco, Egypt, France, China etc. and besides the well-known leafy odour, can have more or less pronounced rosy and minty notes.

Moving on to the fruits, the citrus oils are most abundant — orange, lemon, lime, grapefruit, bergamot. Other oils obtained from the fruit of the plant are *litsea cubeba*, juniper berry and pimento (also from the leaves), and, specifically from the seeds, anise, angelica seed oil and ambrette seed oil. The citrus oils hardly need description, but *litsea cubeba*, with its major component of citral, has a fruity-sweet lemon odour. Juniper berry is green and herbal (the odour blends well with a slice of lemon!) and pimento is balsamic, spicy and peppery. Ambrette seed oil is floral and musky and has fixative properties.

From roots come vetivert oil, orris (whose odour only develops after the rhizomes have been dried and stored), ginger (also strictly speaking from rhizomes) and angelica root oil. You will not be surprised to learn that the commonest adjective for odour applied to this group is "earthy", although all have their own different odour characters. Vetivert is woody and balsamic; orris displays a violet character, ginger is gingery and angelica is spicy and peppery.

Some grasses yield fragrant material: citronella and palmarosa, where the dried grass is extracted, and vetivert — mentioned here again as it is in fact a grass. The two main species of citronella share a fresh character, while palmarosa is rosy but dry and with pronounced bread-like notes.

Many herbs are fragrant. Perhaps lavender is the most well known and the most evocative of natural beauty in terms of sight and smell, but there are many others — lavandin, peppermint, rosemary, sage, and so on. Lavandin is a hybrid of true lavender and spike lavender and has its own distinctive camphoraceous note, although this character is shared by rosemary and sage.

Now to woods such as cedarwood, sandalwood and guaiacwood, which can give warmth and body to a fragrance, and the harsher pine oil and tarry, smoky birch tar. Sometimes only the bark is used; cinnamon bark oil is spicy and sweet and can give an oriental character to a perfume.

Then to some resins and gums that can be extracted from the sap of some plants and trees — galbanum and olibanum resinoids and copaiba and tolu balsams, which can confer fixative properties to a perfume. Galbanum is very green and balsamic; olibanum is very reminiscent of incense.

Finally to a group of lichens that grow on trees and yield oakmoss and treemoss, which are again valued both for their odour and their ability to "fix" a perfume.

As well as the "whole" extract or oil, including the corresponding absolutes and concretes to which I have tended to refer, other parts or fractions — which are still "natural" — are often obtained. These can range from an oil with a specific fraction removed — such as bergaptene-free bergamot oil — to an almost pure chemical species, otherwise known as an "isolate" — linalol distilled from rosewood oil is an example.

Isolates are normally more expensive than the same material produced synthetically — often around five times more expensive — but there are exceptions. It is for example not cost-effective to synthesize eugenol, which is almost always obtained by distillation from clove oil; and some sesquiterpenes, such as patchouli alcohol, are impossible to synthesize.

However, whatever advantages synthetic countertypes may have in terms of price and availability are normally offset in odour terms by their very purity. The trace components in natural isolates give the "rounded", fuller character to their odour which was mentioned earlier and are of great value in higher quality/more expensive perfumes.

For a truly natural fragrance the fragrant material must of course be obtained from the plant by purely physical methods. Those commonly in use today are:

> distillation with or without steam
> extraction by solvent
> expression — from fruit peels

Natural extracts are often further processed and purified using highly specialized techniques which can yield materials with a very consistent odour value and physical characteristics.

3. Synthetic aroma chemicals

These may be produced from natural raw materials by chemical processes, or from organic chemicals derived from turpentine or petroleum, or even from as basic a starting point as acetylene.

Their primary source materials are of course "natural" and, in the case of oils such as pine oil, fairly readily renewable, but whilst they often duplicate materials found in nature, some of them are unique in structure and odour. Around 3,000 aroma chemicals are commercially available although only a few hundred are used in large quantities.

"Natural" fragrances

Returning now to the definition of a natural fragrance, although purists would defend the exclusion of any chemically produced or chemically modified ingredients, you can see why a perfumer will seek to widen it to include at least the synthetic counterparts of naturally occurring substances. There may be more merit in this wider definition, which must in any case exclude chemical species not found in nature, than in a request for a fragrance with only a specified percentage of natural materials, which therefore could include totally synthetic chemicals. A fairly common request today is for a "natural" fragrance containing at least 50 per cent of natural materials with the remainder being nature-identical, however the latter may have been obtained. It is for the end-product manufacturer, the marketer, and perhaps ultimately the consumer, to decide what is acceptable.

If we allow nature-identical materials, the perfumer can use for instance not only eugenol, which is very likely to have been distilled or extracted from clove stem or clove leaf oil, but also isoeugenol, which is normally made by the isomerization of "natural" eugenol but which also occurs in nature in ylang-ylang and nutmeg oils. But should he use it in a fragrance formulation that claims to be natural, at a higher percentage than the maximum at which it occurs in nature? As another example, the damascones occur in rose oil at parts per million. Should the perfumer be allowed to use them at one per cent in a natural fragrance, which he may wish to do, perhaps to overcome a product base odour?

Of more significance, whichever definition is chosen, is the practice of "enhancing" natural oils. We, the fragrance houses, must be on our guard as many natural oils offered for sale are not all they seem! For instance, patchouli oil is often "extended" with benzyl alcohol, fairly easy to detect if you have the necessary equipment and expertise. But some extra synthetic linalol in lavender oil may be much more difficult to detect. There are at least three synthetic routes to linalol and all must be eliminated before you can assume that naturally derived linalol has been used. Most synthetic linalol is produced from acetylene, which gives rise to dihydrolinalol as a by-product; linalol produced from turpentine will contain α-linalol as an impurity, and so on. Gas chromatography analysis under appropriate conditions is essential on any suspect deliveries.

Whilst expressing a preference for total freedom in the choice of raw materials, the perfumer readily appreciates the commercial reasons for requests for natural fragrances but will want you to be aware of the constraints and difficulties he will encounter. Then he (or she) will produce for you the best fragrance which money allows! Natural fragrances do not tend to be cheap because of the scarcity of many oils today and this in turn leads to a plea for more time to answer a request for a natural fragrance, reflecting the greater creative challenge.

The legal position

At the moment there is not one! In the United Kingdom at least, there are no legal definitions of the word "natural" as applied to fragrance. In fact there are generally no legislative restrictions placed upon the fragrance industry, apart from the general requirements not to misrepresent, or to offer for sale anything which is potentially harmful to the user or unfit for its intended purpose. This is due in large part to the efforts of the fragrance industry itself to regulate the use of materials which have been found to be harmful. The International Fragrance Association (IFRA) issues "guidelines" which all of its national members (representing most fragrance suppliers) agree to obey. To date,* IFRA has banned 37 materials previously used in perfumery and restricted the use of 43 others on safety grounds. It is interesting to note that around 30 per cent of these materials are natural, and

---

*This paper was originally presented in 1989. — Eds.

a further 30 per cent occur in nature although they may now be produced synthetically. So natural materials are not intrinsically safer. Hardly surprising when one thinks of some other natural substances — curare, strychnine, poison ivy.....

It is to be hoped that once again the industry can form a consensus on the definition of a natural fragrance rather than have it imposed by legislation. Our parallel industries of flavour and food are currently agonizing over the European Economic Community flavours directive; perhaps we can avoid that, and in fact IFRA is seeking to publish a definition of natural fragrances which will be acceptable to all parties and which will probably be based on the AFNOR (*Association française de Normalisation*, the French Standards Association) definition of natural materials and processes.

Any decision by the fragrance manufacturers will be worthless unless it meets with the approval of the user industries, and ultimately the consumer. Whilst IFRA is very much aware of this, now is the time for users to decide what is practical and to make their views known.

Should any nonodorous components of a perfume, for instance solvents, be excluded when declaring it to be natural? This is the case in flavours and certainly perfumery's preferred solvents do not occur in nature.

Should the words *natural fragrance* be applied only to blends of materials directly obtained from natural sources? Or should the definition be widened to include materials that are known to exist in nature, but may have been produced by chemical modification of natural raw materials or synthesized from alternative chemical species?

Even this definition is still a compromise, of course. The perfumer will be denied his non-naturally occurring synthetics, and with them some fantasy notes and tenacity, but he will be able to provide safe, relatively inexpensive and unique, i.e. creative, fragrances.

As I have said, there are very few totally natural fragrances on the market today, and many which have strong natural "images" are outside both of these definitions. Perhaps a phrase such as "with the natural fragrance of X" is more appropriate.

Natural fragrances are a "new" challenge for the industry but one which we are happy to accept.

Manufacture and Use of Plant Extracts

Keith Helliwell

From the start of his existence, man has searched through the natural kingdom for those plants that might be a source of food. Since he had to rely solely upon his five senses, some of these early adventures undoubtedly proved fatal. However, from his searching arose an armoury of plants with uses other than that of food. Amongst these nonfood plants were those that produced a pronounced pharmacological action on the body: pain killers such as willow bark (*Salix alba* L. and related species) containing an aspirin-like compound, and purges such as senna (*Cassia senna* L.) and cascara (*Rhamnus purshiana* D.C.). Also, there must have been plants used to produce what was considered a beautifying effect, whether to colour the skin or hair (e.g. henna — *Lawsonia inermis*), to soften the skin (e.g. a variety of natural fixed oils such as olive oil) or to produce a change in bodily functions (e.g. the juice from the fruits of belladonna [beautiful lady] — *Atropa belladonna* L., squeezed into the eyes to produce dilatation of the pupils).

Thus, there is a long tradition for the use of plant material, in various forms, in toiletries and cosmetics. At its simplest this would involve the collection of fresh plant material, and either its direct application to the desired part of the body, or the preparation of a simple infusion which could then be used in a variety of ways, thereby increasing the plant's versatility. No doubt fresh plant material was preferred. However, in temperate climates, where fresh plant material was not always available, carefully dried plant material would be substituted, with every effort being made to retain as nearly as possible the supposedly beneficial attributes of the fresh plant. In addition to simple extraction methods, such as the preparation of infusions, more selective methods would be used, such as the distillation of volatile oils and waters and the pressing of plant material for fixed oils, some volatile oils and juices.

Before the advent of modern science with its sophisticated techniques of analysis, man was uncertain as to which constituents of a plant were producing a beneficial effect; thus, he applied to himself as complete an extract of his chosen plant material as it was possible to obtain. Today, in many cases, we can choose an individual constituent, e.g. allantoin from comfrey (*Symphytum officinale* L.), or use a total extract of the plant, e.g. an aqueous/propylene glycol extract of comfrey root, in the knowledge that there may be other constituents in the total extract that are themselves having a beneficial effect.

Many of today's proposed uses of plant materials in toiletries and cosmetics originate from the traditional uses of these plants and their extracts. It would seem reasonable, therefore, to incorporate the plant material into the product in a form relating to its traditional use. In most cases, on a commercial scale, it is not possible to use fresh plant material, apart from fruits, and thus carefully dried plant material has to be chosen. It is essential that the quality of the plant material is adequately controlled, for it is all too easy to fail to identify poorly dried material, and even worse, to use the wrong root or bark because of inadequate expertise in its identification.

Total plant extracts

For total extraction methods, a solvent is chosen which will dissolve from the plant tissue all the soluble components within that solvent's polarity range. Thus, a water-based solvent will not extract non-polar materials such as carotenes and chlorophyll, whereas a solvent such as isopropyl myristate will extract such components. The route chosen for the extraction should be that which retains as much as possible of the intrinsic characteristics of the plant material. Thus, an extract from rosemary (*Rosmarinus officinalis* L.) should smell of rosemary, that from Roman chamomile (*Anthemis nobilis* L.) should smell of chamomile. Plant material that has been too harshly treated during its extraction by heating, poor manufacturing practice or extensive dilution will not measure up to the required standards.

It is essential that the solvent used for extraction fulfils the requirements for the utilisation of the plant extract. Thus, it would be inappropriate to incorporate a water/propylene glycol extract into a hair spray where the solvent would leave an unwanted residue on the hair. An alcoholic extract would be preferable in such a preparation. The usual range of solvent bases is given in table 1, together with some of the applications for each type of extract.

Table 1. Extraction solvents — examples of types of preparations into which the extract could be incorporated

Water	Toilet water, eye gels, shampoos
Water/propylene glycol	Hair conditioners, foam baths, body and hand lotions
Ethanol/propanol	Hair sprays, after-shaves
Isopropyl myristate/fixed oils	Bath oils, body rubs, aromatherapy preparations

The plant equivalent of the extract should also be readily available from the plant extract manufacturer. This information will give the weight of plant material, either fresh or dry, which was extracted to produce a given quantity of extract. It is all too common for arbitrary figures to be quoted. Ask your supplier to explain and justify the basis of his equivalents. It is only by having an accurate equivalent that a manufacturer of toiletries and cosmetics can decide on an incorporation rate of plant material into his product. If the manufacturer is adding a plant extract solely for marketing purposes, the quality and quantity of the added extract are somewhat irrelevant and the old anecdotes about weak, watery soups might well be applicable. However, the reputable manufacturer will realize that the addition of plant extracts should be more than a mere marketing ploy and that such extracts can enhance that product's performance. Thus, it makes sense to incorporate a tannin-containing extract, with astringent properties, (such as birch bark — *Betula spp.*) into an after-shave or an extract with soothing and healing properties (such as marigold — *Calendula officinalis* L.) into a hand cream. However, in order to enhance the product's performance the plant material needs to be present in efficacious quantities, and this is where the plant equivalent of the chosen extract is essential.

It is interesting to note that in products other than toiletries and cosmetics that come into contact with the skin, there is also an awareness of the potential benefit of incorporating plant extracts. One recent example is the launch of a washing-up liquid containing extracts of Roman chamomile (*Anthemis nobilis* L.) and marigold (*Calendula officianalis* L.) to counteract the harsh effect that the act of washing-up has on the hands.

Total plant extracts do suffer from one disadvantage, which some might consider to be an unsurmountable problem. Extracts prepared from roots, leaves, barks and flowers are dark in colour, since in the majority of extracts the plant material from which they are produced is usually dried. The colour ranges from green, if the chosen solvent extracts the chlorophyll from the plant, through browns and oranges to purples and reds when flowers containing anthocyanin pigments are extracted. The incorporation of reasonable levels of the plant extract into the product therefore has a significant colouring effect. This can be offset by choosing a complementary colour for the product and can also be negated by a greater awareness on the part of the consumer that the vibrant colours previously accepted for toiletry and cosmetic products are no longer essential; much as the public has accepted a reduction in the brilliance of the colour of soft drinks following the removal of azo-dyes. The consumer is at present very amenable to alterations to products if he or she can be assured of the product's more "natural" regime.

The major exceptions to the rule that total plant extracts are dark in colour are those extracts prepared from fresh plant material, principally from fruits and seeds. Several of these, as, for example, the extracts of cucumber (*Cucumis sativus* L.), melon (*Cucumis melo*) and lemon (*Citrus limon*), can be virtually colourless, whereas others take on the colour associated with the fruit/seed, as, for example, extracts of apricot (*Prunus armeniaca* L.), horse chestnut (*Aesculus hippocastanum*) and passion fruit (*Passiflora quadrangularis* L.). Examples of nonfruit/seed-based fresh total extracts can be found in lettuce (*Lactuca sativa* L.) and watercress (*Rorippa nasturtium-aquaticum* [L.] Hayek). Since many of these extracts have a low colour intensity, it is possible to offer a range of extracts for any given plant material with differing plant equivalents. The advantage of this is that, with the more concentrated extracts, the plant content not only exerts its claimed effect, but also contributes to the organoleptic qualities of the product into which it is incorporated. Thus, a cucumber eye gel formulated with an extract containing a high plant equivalent of cucumber (*Cucumis sativis* L.) would not only offer the beneficial effects of the cucumber but would also contribute a mild cucumber odour originating from the extract itself, obviating the need for artificial fragrances.

Total plant extracts have been restricted, in the past, to liquid preparations which have been incorporated into mainly liquid and semi-solid products. Recently work has been undertaken to produce powdered total plant extracts on a selection of inert bases. Those produced with a talc base have applications in talcum powders and a wide range of colour cosmetics. Those produced with a sesquicarbonate base have applications in various bath preparations. Although no products based on this concept are currently on the market, several prototypes are undergoing trials and it is anticipated that in the not too distant future commercial products will be available.

Partial plant extracts

In the previous section we have looked at total plant extracts, that is, those that are prepared either by maceration or by percolation of one or more of the organs of the plant (root, leaf, flower, etc.) in a range of solvents. Partial plant extracts, on the other hand, are produced by fractionating a particular constituent or range of constituents by either chemical or physical methods from one or more of the organs of the plant. This may be an oil (fixed, volatile, absolute, concrete, etc), a juice or a sap. The juice of a citrus fruit might be expressed to produce a partial plant extract, but this would be different in composition from the total extract where the whole fruit is first comminuted and then macerated in a suitable solvent.

Fruit juices, in varying degrees of concentration, are widely used in certain preparations, particularly those for addition to the bath or shower and those to be applied as lotions to various parts of the body. The type of product determines the quality of the juice: an opaque product, such as a cream or lotion, does not require a high clarity juice, and certain products on the market are even amenable to the incorporation of the comminuted whole fruit rather than expressed juice. However, a clear product requires a high-clarity juice and great care needs to be taken to remove unwanted plant debris.

The saps or exudates from various plants are another group of partial extracts. The extract from *Aloe vera* falls into this category, as do various balsams and gum resins such as tolu (from *Myroxylon balsamum* [L.] Harms), benzoin (from *Styrax tonkinensis* [Pierre] Craib ex Hartwich) and styrax (from *Liquidamber orientalis* Miller). Their extraction and initial processing take place, of necessity, in the area where the plants are cultivated. In many cases, the semiprocessed plant extract is offered onto the market in different grades.

The extraction of fixed oils, volatile oils, absolutes and concretes falls outside the scope of this article. However, one aspect of their production and use — the production of aromatic waters — is worthy of consideration. Aromatic waters may be classified as follows: (i) distilled waters; (ii) prepared waters and (iii) concentrated waters.

i) Distilled waters

Distilled waters can be considered as true partial extracts of the plant material from which they are derived and may be by-products of the distillation of volatile oils, e.g. chamomile water and lavender water, or may be produced by direct distillation of plant material, no oil being collected, e.g. elderflower water. Their method of preparation, by definition, necessitates that they be totally natural products with no added synthetic fragrance components.

ii) Prepared waters

Prepared waters are produced by dissolving volatile oils, concretes or absolutes in water to produce a saturated solution. These may be produced solely from the natural partial plant extract (volatile oil, absolute or concrete) or they may be wholly or partially synthetic. The aromatic waters produced by this method that have equivalents in category (i) above are missing the top fresh notes of their category (i) counterparts.

iii) Concentrated waters

Concentrated waters invariably contain an alcohol which is present to maintain in solution high levels of oil. As with category (ii) above, the volatile oil, absolute or concrete, may be wholly or partially synthetic. Concentrated waters are often diluted on the basis of 1 part of concentrated water to 39 parts of water to produce an aromatic water. The aromatic water so produced is not necessarily equivalent in either strength or odour to the corresponding prepared water (ii above), made by dissolving the volatile oil, absolute or concrete, in water to produce a saturated solution.

Aromatic waters, as with other partial plant extracts, will only possess the claimed activity of the plant if the active constituents of the plant are contained within that fraction of the plant which forms the partial plant extract. Many partial plant extracts, particularly distillates, have the advantage of being virtually colourless, making for ease of incorporation into a range of products. Unfortunately, this has led over the years to the production of distillates from a wide range of plant materials which are not suitable to this form of processing, and therefore little credence can be given to any claimed activity of the plant material or its extract.

Table 2. Applications and uses for a selection

	A	B	C	D	E	F	G	H	I	J	K	L
Anthemis		x	x		x	x	x	x	x	x	x	x
Apricot				x	x		x		x			x
Arnica	x	x	x	x								
Birch (bark)	x	x	x	x		x						x
Calendula	x	x	x	x	x	x	x	x	x	x	x	x
Comfrey (root)	x			x		x	x	x			x	x
Cucumber				x	x	x	x		x	x		x
Equisetum	x	x	x	x	x	x		x		x	x	x
Horse Chestnut												x
Lavender	x	x	x		x	x		x	x		x	x
Lime Flowers	x		x	x		x	x	x		x	x	x
Marshmallow		x	x	x		x	x		x	x		x
Matricaria						x			x			
Meadowsweet				x		x		x		x		x
Nettle	x	x	x			x		x	x			x
Papaya				x	x		x		x			x
Rosemary	x	x	x		x	x	x	x			x	x
Sage	x	x	x	x	x	x		x	x		x	x
Watercress	x				x				x		x	x
Yarrow	x	x	x		x	x		x	x		x	x

Key:

A.	HairTonic/Rinse	H.	Astringent/Toner
B.	Shampoo	I.	Face Preparation
C.	Conditioner	J.	Eye Preparation
D.	Cream	K.	Foot Preparation
E.	Lotion	L.	Bath Preparation
F.	Soap/Cleanser	M.	Normal Skin
G.	Moisturizer	N.	Dry Skin

of commonly used plant materials

M	N	O	P	Q	R	S	T	U	V	W	X	Y	Z	a	b
x		x		x	x	x				x	x				
x		x	x												
x	x														
						x			x	x					
x	x	x				x			x		x		x		
x		x	x			x		x		x					
x	x	x		x	x	x									
	x			x	x	x	x	x	x						
	x				x				x	x		x			
x				x		x						x			
x				x	x	x		x							
					x	x									
				x					x	x	x				
x	x														
	x	x	x	x			x		x	x		x			
	x				x		x	x		x		x	x	x	x
					x	x									
	x	x	x		x	x			x	x					

O.	Oily Skin
P.	Sun-screen
Q.	After-sun
R.	Ageing/Antiwrinkle
S.	Oral
T.	Sensitive Skin
U.	Normal Hair
V.	Dry Hair
W.	Oily Hair
X.	Antidandruff
Y.	Light Hair
Z.	Dark Hair
a.	Auburn Hair
b.	Brunette Hair

Technical aspects related to the manufacture and use of plant extracts

In order to define a plant extract, the following information should be readily available:

(a) a description of the plant material used for the production of an extract; this should include both a botanical description (genus and species) and an indication of which organ (root, leaf, flower) was extracted

(b) the solvents used in processing the extract

(c) the plant equivalent (see above under "Total plant extracts"); it should also be possible to provide a plant equivalent for a partial plant extract

(d) whether there are any additives to the extract, e.g. preservatives, colours, fragrances

(e) an analytical specification with which the extract should comply — the physical/chemical characteristics to be measured will vary according to the extract type

(f) a statement on maximum levels of microbiological levels of contamination

(g) a statement on the shelf-life of the extract

(h) an indication of storage conditions and any precautions to be taken in handling the extract

Much information is available from the literature concerning the toiletry/cosmetic applications of various plants. Some commonly used plants have been selected and an applications chart for these is given as table 2 (above). In addition, producers of plant extracts often have trial formulations available to help evaluate the performance of an extract in a certain product type. Thus, for the innovator of new toiletry and cosmetic products there need be few restrictions on the usage of plant materials in his/her new trials. A reputable manufacturer of plant extracts should be able to tailor-make either a single- or multiple-component plant extract to the needs of the product both from the point of view of the correct solvent for compatibility with the product and of achieving the required levels of incorporation of the plant material. However, it is essential that both the product designer and extract producer collaborate fully from the outset of any project.

Acknowledgements:
I would like to thank my colleagues J.A. Whitehead and A.W.G. Esmond for collating the information for the applications chart.

Preparation of Herbal Extracts and Their Effectiveness*

Peter Jarvis

1. Sources

In Europe we tend to regard herbal cosmetics as an invention of the late twentieth century and, indeed, it was not until around 1965 that we became aware of them through the famous bath product "Badedas", which used extract of horse chestnut, presumably for its cleansing effect (via saponins). Modern advertising implies, however, that the extract has far more magical properties.

A limited variety of extracts other than horse chestnut were also available at that time and were being used in cosmetics, but in a fairly small way and then mainly on the Continent, particularly in France and Germany. They were all well-known herbs, e.g. rosemary, chamomile, fennel, pine, birch and cucumber; furthermore all were of European origin. Since then, however, and probably as a result of the success of "Badedas", herbal products have developed considerably. Cosmetic chemists began not only to consider other common (and some not-so-common) herbs for use in their products, but to look beyond Europe for their herbs. Henna (from Egypt, Persia or India) and ginseng (from Korea) were two of the first of the non-European herbs to become popular.

This, however, was not really the start of herbal cosmetics — only the start of how we think of them today. Herbs have been used in medicine and skin care — taken internally or applied externally — since the start of civilization and many ancient herbals, e.g. , Culpeper's, having lain untouched for many years

---

*Originally entitled "Herbal Extracts — The Truth".

after the advent of modern medicine and drugs, are now being avidly researched for even the remotest reference to a particular herb's use in skin preparations.

Many of the European herbs we use for our processing nowadays are from Eastern Europe, where the climate (and presumably labour costs) favours their production, but there are many other producing countries. England itself provides a few, notably chamomile and elder flower. Some extracts are required in such small quantities, e.g. brooklime, that we cannot use a commercial supplier of the dried herb for processing and have to do the gathering ourselves. This requires the cooperation of some friendly farmers who allow the herb to grow in stretches of their private water meadows. I emphasize *private* since we do not wish to be accused of interfering with wild flowers, most of which are protected. I know of no law, however, that prevents one from harvesting wild flowers if they are grown in a "commercial", albeit informal, way on private property.

The particular plant organs that are used for extraction vary from herb to herb and in some cases separate extracts are made from different organs since they contain entirely different compounds, e.g. elder flower and elder berry. The particular organ chosen is usually the part recommended by the early herbalists for its medicinal value and so continues to be the most harvested and thus the most easily available.

A study of the more modern herbals, and Mrs Grieve's is probably the most useful in the English language, reveals a wide range of chemical bodies present in each herb. Whilst I am not decrying this great work, it should be pointed out that it is made up of a meticulous literature search from other sources and most of the chemical compositions are, I believe, taken from Steinmetz' *Codex vegetabilis*, which lists all the compounds identified in each herb.

Chemical techniques were, however, not as exact in those days as they are today and I suspect that many of the chemical entities found and attributed specifically to one herb may in fact be identical to those that were thought, from early analytical methods, to be specific to other herbs.

For extraction purposes, dried herbs are invariably the form used. The reason is fairly obvious: dried herbs are not subject to fermentation and thus can be stored until required.

It is quite possible that during the drying stage and in storage afterwards many chemical changes take place, but for us this is of no great importance. Our main concern is that the extract has been obtained from the herb named, not whether it was fresh or dry.

2. Production of extracts

The first stage is the production of a pre-extract. This consists of leaching out the usable part of the herb by a solvent in order to eliminate the unusable insoluble matter such as cellulose and lignin from the cell wall.

Whether dry or fresh herbs are used, this process is preceded by a comminution stage to break the herb down into small particles, thereby rupturing the cell walls so that the maximum surface of herbal matter is presented to the solvent.

The comminuted herb is then treated with a solvent and this is performed simply by stirring over a prolonged period or by percolating the solvent through the herb suspended in a basket. A more complete extraction is obtained if the solvent is at elevated temperatures and, if the solvent is a volatile one, it is obvious that this can be performed in a Soxhlet-type of extractor in which the solvent is evaporated, condensed and allowed to flow over the herb whence it falls into the boiler to have the solvent re-evaporated to repeat the cycle.

The choice of solvent will depend entirely on what one wishes to extract from the herb, and whilst the commonest solvent, i.e. water, will take out sugars, saponins and mineral salts, ethanol will extract chlorophyll, essential oils (where present), some fatty materials and other organic materials.

This pre-extract is then clarified by centrifuging or filtering and concentrated by evaporation of some of the solvent, when a dark-coloured syrupy liquid results.

Alternatively, it may be taken to a dry stage by spray-drying, usually with a simple sugar or an inorganic phosphate. Generally speaking, preservatives are not added to these pre-extracts since the concentration of dissolved solids, and hence the osmotic pressure, is high enough to render this unnecessary.

We mentioned that these pre-extracts are usually dark-coloured syrupy liquids. Why this is so we do not know. Certainly tannins are responsible in some cases but in others they are not, and we know of no explanation for the colour in non-tannin-containing extracts.

Finished herbal extracts as we know them, for use in cosmetics, are prepared by dilution of the above-mentioned pre-extract using a solvent followed by clarification and possibly the addition of a preservative.

One might question the need for further clarification since the pre-extract has already been clarified, but the choice of diluent will often make this necessary. The three most usual diluents are

      i   water
      ii   glycol/water mixtures
      iii ethanol/water mixtures

and all have their pros and cons. Suppose we start with a pre-extract which has been prepared using water as the extracting solvent.

i. Water will be a perfectly good diluent and will produce an extract compatible with all cosmetic preparations except those high in ethanol content or preparations consisting wholly of oils. Such an extract must, however, be preserved if it is to have a reasonable shelf-life.

ii. Glycol/water will suit most products, including alcoholic ones, since glycol precipitates many of the plant materials which alcohol precipitates and these will have been eliminated by the clarification stage.

iii. Ethanol/water will be the more useful diluent if the extract is intended for use in an alcoholic preparation since the clarification following dilution will eliminate the ethanol insoluble matter.

It will be obvious from the above, however, that since precipitation occurs with glycols and ethanol at the preclarification dilution stage, one is removing some of the herbal matter and it is not just a small proportion. One can see on making the dilution that sometimes in excess of 50 per cent of the dissolved herbal solids is removed in this way. Thus, it can be more economical when one wishes to introduce a fixed amount of herbal solids to a preparation to use, say, 0.5 per cent of aqueous extract rather than 1 per cent of a glycolic one.

The calculation of plant solids is the main guide to the extract's strength and shows how much herb you are getting for your money. Thus the most reliable way to estimate the strength of an extract is by drying an accurately weighed quantity to constant weight and calculating the solids content. It is most important, however, in the case of a glycolic extract, to ensure that all the glycol has in fact been evaporated and this is a lengthy process. Many people have been misled into thinking that they have an extract with perhaps 30 per cent of plant matter when, in fact, they have failed to evaporate the glycol and are counting this as plant matter.

Hydro/alcoholic extracts are used in alcoholic preparations when the use of an aqueous extract could lead to precipitation of some of the plant material by the alcohol. However, most alcoholic preparations are filtered nowadays, so there is no particular advantage in not using the aqueous extract and filtering out the resulting precipitate. The price advantage could be significant also.

3. Method of incorporation

In aqueous or aqueous/alcoholic preparations the extract may be incorporated at any stage of manufacture. In emulsions it is usual to incorporate it along with the perfume, i.e. during the latter stages of cooling. In dry products like soap and talc the extract is incorporated at the same time as the perfume and in the same way.

We would emphasize to developers of cosmetics that high pH values invariably cause a darkening of the extract, a factor which should be taken into account especially if the product is a delicately coloured one. Thus it is often better during development stages to work with an extract that has been made slightly alkaline. Such will be the case for a Carbopol gel when, if this has not been done, the final neutralization stage may produce a marked colour change, and the work would then need to be repeated with a lesser extract content.

4. Claims

We believe that if one had sufficient time to research all the available literature, one could find a reference of efficacy towards any minor skin ailment for any herb that can be named. During the past few years, the market has been flooded with books about herbs and their effects. The best we have found are *A Modern Herbal* by Mrs Grieve, but *Guia de las plantas medicinalis* by Paul Schauenberg and Ferdinand Paris and Steinmetz' *Codex vegetabilis* are also most useful reference sources.

Traditionalists will perhaps favour the older herbals like Culpeper's, but they are ponderous to read and tend to claim cures for everything under the sun including such obscure ailments as "wens and kernels" and "king's evil". Mrs Grieve has done us all a favour by separating the wheat from the chaff of these ancient herbals.

Let us suppose, however, that one wished to manufacture a moisturizer containing a certain herb and wished to justify its inclusion in the product. A detailed literature search might yield that the herb "soothes and protects the skin". How is one to react? Is the claim that we have found saying, in an old-fashioned way, that the herb has a moisturizing function?

Ask a technical person whether it is right to use this herb for this claim and he is likely to say no. Ask, on the other hand, the sales and marketing section and they may be only too keen to say yes. After all, the name of the herb might be just what is needed to achieve an attractive product title and they are, after all, interested in selling products against their competitors.

Perhaps the advertising copy writer could come to the rescue. Could he perhaps write some clever copy which, by the use of some "weasel" words such as *helps*, would, without making any specific claims, make the public believe that this herb really is the last word in moisturizing. Almost certainly he could, but this immediately raises a moral question, Is it correct to do so?

Read through any woman's magazine and study carefully the advertisements for skin-care products and more precisely the claims. There will be no shortage of "weasel" words and phrases extolling the virtues of various "wonder" ingredients. I think the marketer is likely to say that if he cannot beat them he must join them and go ahead and make his claim. Either that or lose out to his competitors.

Perhaps the moral question would not prove so heart-searching if the literature references were far more definite in their moisturizing claim. Certainly the technical man would be far happier to give his blessing, but this brings us to the next question of the herb's effectiveness.

5. Effectiveness

Are herbs really effective in achieving the claims made for them? Without doubt some of them have some markedly powerful properties: for instance, curare vine, cactus, poppy, foxglove and coffee, which yield curare, mescalin, morphine, digitalin and caffeine, respectively. Nobody would argue about their dramatic effects, albeit when taken internally, but we are concerned only with topical action when applied via skin-care preparations.

The claims made by the herbals for a herb's topical effects never mention how much is to be applied, but I think we can take it that when they talk about the application of the decoction of a herb, they are talking about a goodly application — far more in fact than is likely to be present in an application of a few grams of cream. Obviously any efficacy would be greatly enhanced by inclusion at higher levels, as is practised by many successful companies.

This raises another moral question. How can we extol the virtues of a material included in our product when it is there in a quantity perhaps insufficient to have any action? Should we, perhaps, face up to the fact that we are to some extent "sellers of dreams"?

6. Future

Is there then a future in selling dreams? The answer is very definitely yes, since, if what we sell makes people feel better, then we are sure it does them a power of good.

We are not saying that good, well formulated functional products do not make the skin feel more comfortable, look better and exert a protective function, but we regard most of the "special" additives as having mainly a placebo value, and it is the promises of clever advertising that make the consumer feel that the special additive in which he or she has faith is the reason why the cosmetic with the "wonder" ingredient is far more effective than the one without.

The market shelves are crammed with excellent skin-care products, all superbly packaged, but all of more or less equal efficacy, and consumers are confused since there is so little to choose between them. One way in which we can make their choice easier is an attractive product title with attractive claims by way of an incorporated wonder ingredient.

The question is, What can the wonder ingredient be?" We all know that so-called "natural" cosmetic preparations are definitely "in" and as the public's attitude grows "greener" by the day, animal products are becoming taboo. Herbs are the obvious answer and are available in plenty. The extracts of at least 200 different herbs are on offer to cosmetic formulators at the moment and the surface has still only been scratched. There still remain to be utilized herbs with interesting properties and attractive names to enhance product titles for decades to come.

Use of Plant Extract Unsaponifiables in Liposome Form*

M. Benhaim, Jean Bommelaer and B. Levrat

I. Introduction

Between 300,000 and 400,000 plant species grow on our planet, but the cosmetic and pharmaceutical industries use only some 2,000. However, many effective cosmetic treatments have been developed using a diversity of aqueous, alcoholic or hydroglycolic plant extracts.

Plants contain common components such as minerals, cellulose, sugars and proteins as well as secondary metabolites such as polyphenols, alkaloids, terpenoids, lipids, vitamins, essential oils and unsaponifiable compounds.

The differences in lipid phase activities in plant extracts are mainly due to the nature of their unsaponifiable content.[1,2] Biochemical studies performed on different fractions of plant extracts suggest that unsaponifiables have a particularly valuable pharmacological action.

It has been established that although the nature of fatty acids is almost the same whatever the vegetable oil may be, these oils can be clearly differentiated by their unsaponifiable fraction.

II. Composition of unsaponifiables from plants

In most vegetable oils, unsaponifiables represent a minor fraction of the total fatty compounds.[3]

Unsaponifiables are generally very complicated mixtures, and the composition of any one plant extract has not been completely

---

*Originally entitled "Utilization of Unsaponifiables from Plant Extracts in the Form of Liposomes".

established. Furthermore, the chemical structure of some compounds is not well defined although their identification has been considerably improved through the development of chromatography and other analytical equipment. The classical unsaponifiable substances are as follows:

sterols, such as cholesterol, campesterol, stigmasterols and
    β-sitosterols
terpenoids
coloured pigments, such as carotene and carotenoids
tocopherols
aliphatic alcohols
hydrocarbons, such as squalane and squalene

III. Methods for extracting unsaponifiables

Various processes are available which enable one class of molecule to be extracted exclusively. A vapour-carrying technique, for example, is used to extract essential oils. The lipid phase is extracted by organic solvents. Molecular distillation or lipid-phase saponification is necessary to isolate the unsaponifiable fraction.

IV. Methods for identifying unsaponifiables

(i) Thin-layer chromatography
Thin-layer chromatography of complicated mixtures produces spots representing different families of compounds.

These spots are identified individually by parallel chromatography of reference substances. Separation quality can be improved by modifying several parameters such as the choice of support, the migration method or the composition of eluant. Thus, sterols, tocopherols and polyunsaturated alkenes can be isolated.

The most commonly used eluting solvents are mixtures of polar and non-polar materials, as follows:

hexane/ethyl acetate
hexane/ethylene oxide
pentane/chloroform

The development and identification of the different bands are made either by UV development or by spraying with special reactants which produce typical colorations. Some reactants, such as iodine, destroy the identified compound and do not allow further identification, but others give a reversible fixation. A simple elution of the developed fraction on a separative column enables the compound to be obtained without reactant.

(ii) Gas chromatography
Gas chromatography allows a precise identification of each component of the spots observed by thin-layer chromatography[5].

High-resolution spectra are obtained on capillary columns (fig. 1). Most of these columns are silicic supports impregnated with variable-composition stationary phases. Gas chromatography is particularly effective when used with a mass spectrometer, which gives a precise molecular identification of each separated substance.

(iii) Liquid chromatography
Liquid chromatography is a suitable method for fatty acids, flavonoids and vitamin analysis but does not give good results for complex mixtures of plant unsaponifiables, which is why it is not commonly used for this kind of analysis. It enables precise dosages to be employed, especially for oestrogens, tocopherols or carotenoids. When the separation is successful, the conditions can be scaled up to a preparative stage.

V. Unsaponifiables in aqueous media

Unsaponifiables from plant extracts are not currently used in cosmetics because of their total insolubility in water and most organic solvents.

Exsymol has developed two methods of solubilizing plant unsaponifiables to enable their use in cosmetic formulations. Unsaponifiables are available either in the form of extracts (Hydrumines) or as liposomes (Exsyliposomes).

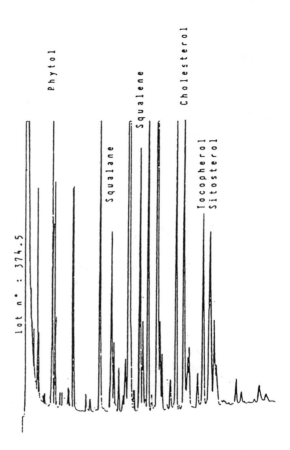

Capillary column: OV 17
Eluting gas: Nitrogen
Oven temperature   180-290°C (2°C/min.)
Injector temperature   275°C
Detector temperature   270°C

Reference substances: *Sitosterol, Cholesterol, Squalene, Phytol, Squalane* and *Tocopherol.*

Figure 1. Identification of unsaponifiable compounds in *Humulus lupulus* by gas chromatography

1. Hydrumines

Hydrumines are total plant extracts in which the water-soluble fraction and the unsaponifiable fraction are present. The water-soluble fraction contained in 1 kg Hydrumines corresponds to the water-soluble fraction of 1 kg of fresh plants. This water-soluble fraction is then enriched with unsaponifiables. The resulting solution corresponds to 1 = 33, meaning that 1 kg of Hydrumines contains the unsaponifiable fraction of 33 kg of fresh plants.

2. Exsyliposomes

2.1 Structure

Exsyliposomes are true unilamellar liposomes, which means they are very small vesicles dispersed in an aqueous phase. Their wall is composed of a natural glycerophospholipid bimolecular layer.[6,7] They are called liposomes because their membrane has the same constitution as cellular membrane (fig. 2). The manufacturing process makes the membranes sufficiently rigid to avoid fusion. The permeability of the membrane is reduced to avoid the active principle escaping from the liposome. The liposome membrane is protected against oxidation. Exsyliposomes imitate cells by enclosing lipophilic fractions of plants in their membrane and the hydrophilic fraction in their centre.

2.2 Characterization

The size of Exsyliposomes is measured by laser light diffusion and by electron microscopy.

2.2.1 Laser light diffusion

A laser beam is passed through the sample, when particles diffuse a certain amount of photons in all directions. The correlation function of the diffused intensity which decreases exponentially is proportional to the diameter of the particles in Brownian motion (fig. 3).

2.2.1.1 Unimodal analysis (Gaussian distribution)

An average value of size, with standard deviation, is obtained. This method sometimes gives unreliable results because, in the

hydrophobic tail

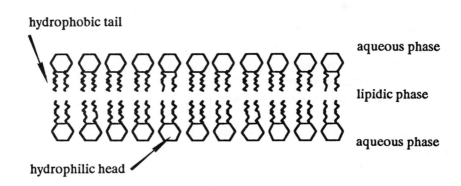

aqueous phase

lipidic phase

aqueous phase

hydrophilic head

Glycerophospholipid membrane

Formation of
closed vesicles

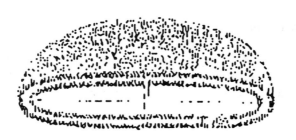

Figure 2. Structure of Exsyliposomes

case of a heterogeneous population in solution, small and large particles will be totalized, and the average size will be given even though a small number of large particles diffuses as much light as a large number of small particles.

2.2.1.2 Bimodal analysis

An adapted computer programme allows polydispersed samples to be dispersed into subpopulations. The relative proportions of each class in intensity, percentage of the total population and weight are also obtained. The technique of varying sample times increases precision.

2.2.2 Electron microscopy

This involves the classic technique of negative coloration by uranyl acetate (fig. 4). Liposomes are subjected to aggressive treatment: heavy metal salts modify the ionic strength of the liposomes. The negative coloured solution is then dried and the liposomes are observed in very high vacuum under a high-energy electron beam.

2.2.3 Results

Both of these methods are complementary and give a reliable image of the solution. The distribution of particle sizes is highly dependent on manufacturing conditions. If perfectly defined, these conditions give high reproducibility in terms of size and population homogeneity. Exsyliposomes of *Centella asiatica*, for example, give the following values:

Table 1. Bimodal analysis of Exsyliposomes of *Centella asiatica*

Lot No.	Size nm	Weight %
J249	27	99
J282	29	99
J283	30	97
J287	29	97

Unimodal values do not provide a reliable indication of the solution state because of diffusion of the large-sized particles. In

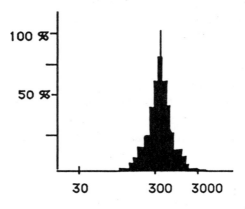

Unimodal analysis

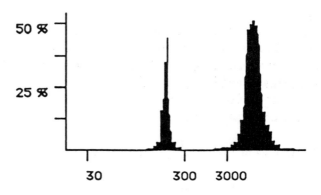

Bimodal analysis

Figure 3. Characterization of Exsyliposomes by laser light diffusion

bimodal distribution, the average size is about 30 nm and represents 98 per cent of the total population. The values obtained by laser diffusion are confirmed by electron microscopy.

In another study it was shown that Exsyliposomes are stable in aqueous solutions several months after manufacture.

2.3 Identification of unsaponifiables

We chose to identify the unsaponifiable fraction in Exsyliposomes by the method of gas chromatography.

It is first necessary to obtain a lipid extract from Exsyliposomes and separate the unsaponifiables from the phospholipids which constitute the membrane (fig. 5). A basic hydrolysis of the phospholipids renders them soluble in the aqueous phase. Unsaponifiable compounds are extracted with an organic solvent, such as diethyl ether, which is evaporated, and the resulting extract dissolved in chloroform for silylation. Silylation involving N,O-*bis*(trimethylsilyl) trifluoroacetamide (BSTFA) gives excellent results in this case. The silylated solution is analysed by gas chromatography on a capillary column. The spectrum obtained corresponds to the characteristic spectrum of the unsaponifiables of the plant used (fig. 6).

Unsaponifiables are thus faithfully transferred from the plant to the liposome. "Liposomation" of unsaponifiables does not alter either their composition or their structure and they can be perfectly identified after extraction from liposomes.

VI. *In vitro* study of activity of unsaponifiables

It is also worthwhile comparing the activity of unsaponifiables in liposome form and in Hydrumine extract form on living material simulated by a cell culture.

1. *Principles and objectives*

An *in vitro* culture of dermic cells (fibroblasts) was prepared in order to prove the cytostimulating or regenerative action of the active principle on cutaneous connective tissue, according to the method of Meynadier.[9] A monolayer culture of human fibroblasts in an incubative medium containing a suboptimal

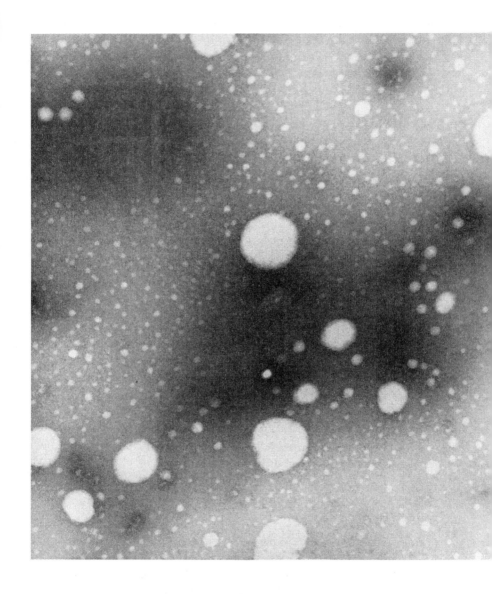

Figure 4. Electron microscopy: structure of Exsyliposomes
(magnification × 60,000)

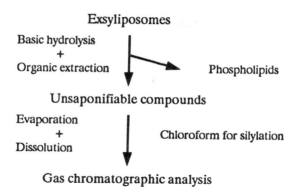

Figure 5. Identification of the unsaponifiable fraction

concentration of foetal calf serum was prepared. Cells were thus placed in a depressed environment, characterized by Adolphe,[10] having a very slow cell multiplication rate. In such conditions, a cytostimulant ingredient is able to increase cellular growth. Thus, we established an artificial model of dermic cells, artificially aged, having old cell characteristics.

This type of experimental model was used to study the cytostimulating and regenerative action of our ingredients on skin, according to the technique of Gospodarowiez et al.[11] Evaluation of cellular proliferation was made by cell numeration.

2. Definition of tested products
The solutions to be tested were incorporated in the culture medium at 0.1 μg/l concentrations, expressed in unsaponifiables.

solution A$_1$: control with 10% foetal calf serum
solution A$_2$: control with 3% foetal calf serum
solution B: containing Hydrumines of *Centella asiatica* Lot No. 514002
solution C: containing Exsyliposomes of *Centella asiatica* Lot No. J281

3. Materials and methods
3.1 Initial cellular culture
Human fibroblasts were isolated from skin fragments free from subcutaneous fatty tissue. The 1 mm^3 fragments were placed in

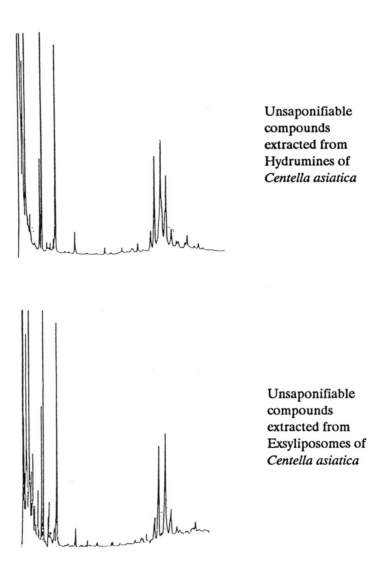

Unsaponifiable
compounds
extracted from
Hydrumines of
Centella asiatica

Unsaponifiable
compounds
extracted from
Exsyliposomes of
Centella asiatica

Figure 6. Comparison of unsaponifiable fractions derived from
Hydrumines of *Centella asiatica* and Exsyliposomes of *Centella asiatica*

culture dishes and covered with 1 ml of Minimum Eagle Medium (MEM), followed by 10 per cent foetal calf serum (FCS), 200 IU/ml of penicillin G, 100 μg/ml of streptomycin and 2 mM of glutamine. Cultures were incubated at 37°C in a humid atmosphere with 95% O_2 and 5% CO_2.

Fibroblasts grow from the explants; they can be subcultured after 2 or 3 weeks. Cells were quantitatively separated from the culture dish after a 5-minute exposure to a trypsin solution, and collected after centrifuging. The cells were suspended in the growth medium and distributed in the culture dishes. Several changeovers were made before using the cells for the cytostimulation study.

3.2 Experimental method
Cells were distributed in a 24-vessel Costar plate apparatus, with 20,000 cells on each vessel, in 0.5 ml of 3% FCS diluted in MEM. As soon as the fibroblasts were fixed, the volume of the medium was adjusted to 1.5 ml per vessel. Solutions to be tested were diluted to selected concentrations in this volume. A reference plate was also prepared (table 2). Cultures were then placed at 37°C in a 95% O_2 and a 5% CO_2 atmosphere.

3.3 Statistical study
The mean, statistical error and standard deviation were calculated for each lot, both reference and treated. A statistical comparison was made according to Student's *t*-test.

4 Results
Cellular growth was increased significantly (p <0.05) when unsaponifiables were added to the culture medium at the level of 10 μg/ml. Cellular growth activation with unsaponifiables in the form of extracts or liposomes was 33 per cent.

Per cent activation: $$\frac{M_E - M_{3\%}}{M_{10\%} - M_{3\%}} \times 100$$

M_E = cell number for the tested product.
$M_{3\%}$ = cell number for 3% FCS control (A2 solution).
$M_{10\%}$ = cell number for 10% FCS control (A1 solution).

Table 2. *In vitro* study experimental method

Material	Human fibroblasts		
Incubation medium	Dubelco MEM		
	10% FCS	3% FCS	
Treatment		Hydrumines of *Centella asiatica* 10 mg/l in unsaponifiables	Exsyliposomes of *Centella asiatica* 10 mg/l in unsaponifiables
Dosage	Cell numeration		

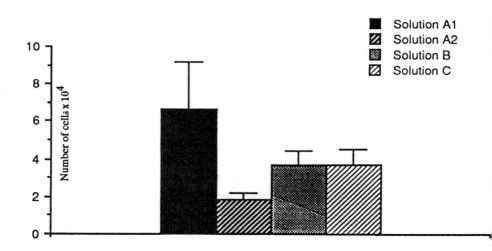

Figure 7. Action of *Centella asiatica* unsaponifiables
on fibroblast proliferation

VII. Conclusion

The influence of cosmetic active products on the proliferative ability of human dermic cells was studied. A fibroblast culture in suboptimal growth conditions was used to estimate the cytostimulating activity of a regenerative ingredient on an ageing dermis. Fibroblast cult ures usually proliferate in a culture medium enriched with 10 per cent FCS; in these conditions growth is optimal.

Suboptimal growth is obtained by culturing cells in a medium partly lacking mitogenic factors, which are usually present in serums. A medium containing only 3 per cent serum was therefore used and the *Cente lla asiatica* unsaponifiables added to the 3 per cent FCS mediu.n increased cellular growth. Growth level differences between the control (3% FCS) and the sample treated with unsaponifiables became significant ($p < 0.05$).

It can be seen that the liposome form does not provide any synergy compared with the cytostimulating activity of *Centella asiatica* unsaponifiables at a level of 10 mg/l of unsaponifiables. However, these liposomes are still of interest because they enable cosmetic chemists to incorporate unsaponifiable compounds in aqueous phases without the use of a surfactant. Furthermore, the described technique is useful for the study of potentially mitogenic substances, such as plant unsaponifiables which are able to increase cellular multiplication and connective tissue growth.

References

1. Expansciences, French Patent, FR 2,041,594, 1971.
2. Wolf, J.P. and Walbeck, W. "Les constituants de l'insaponifiable". *Jnées Inf. ITERG,* Marseilles, 27, 1963.
3. Rancurel, A. "L'avocat: son huile et son insaponifiable. Utilisation en cosmétique". *Parfums, Cosmétiques et Arômes,* N° 61, 91-95, 1985.
4. Mordret, F. "Applicaltion de la chromatographie en couche mince à l'étude de quelques constituants insaponifiables". *Bull. Soc. Chim. France,* 1, 84, 1973.

5. Mordret, F. *Revue Française des Corps Gras,* 16ème année, N° 10, p. 639-652, 1969.

6. Puisieux, F. and Delattre, J. "Les liposomes, applications thérapeutiques". Ed. Lavoisier Tec. & Doc., 1985.

7. Knight, C.G. *Liposomes: From Physical Structure to Therapeutic Applications.* Elsevier/North-Holland Biomedical Press, 1981.

8. Cummins, H.Z. and Pusey, P.N. "Dynamics of macromolecular motion". *Photon Correlation Spectroscopy and Velocimetry.* Plenum Press, 1979.

9. Menadier, J. *Précis de Physiologie Cutanée.* Ed. Porte Verte, Paris, 1980.

10. Adolphe, M., Pointet, Y., Ronot X. and Wepierre, J. *Int. J. of Cosmet. Sci.,* 6, 55-58, 1984.

11. Gospodarowiez, D., Greenburg, D., Bialecky, H. and Zetter, B.R. *In Vitro.* 14, 1978.

Analysis of the Claimed Efficacy of Chamomile*

Anthony C. Dweck

Synopsis

This paper analyses the frequency and sources of the benefits attributed to chamomile and illustrates the widely varying claims made for this herb according to the source consulted. It highlights the importance of a thorough literature survey before making product claims.

There is a moral and ethical obligation for all of us to try to justify what we put on our packs, while our marketing teams are duty-bound to make the product as tempting as possible to the consumer. An honest balance must be achieved between the claimed benefit, the dosage, the quality and the substantiating evidence.

Chamomile was chosen as an example of a well-documented herb to illustrate the problems.

The first hurdle is to select one of the many common names. *Anthemis nobilis,* or *Chamaemelum nobile,* is know as Common Chamomile, Spanish Chamomile, Belgian Chamomile, English Chamomile, Russian Chamomile, Garden Chamomile, Low Chamomile, Double Chamomile, Ground Apple or Whig Plant.

Matricaria chamomilla is not a true chamomile and is more correctly called *Matricaria recutita.* The common names are German Chamomile, Sweet False Chamomile, Hungarian Chamomile, Single Chamomile, Scented Mayweed and Wild Chamomile.

---

* Originally entitled "Claims for Herbs and Extracts — A Search for the Truth".

There are other members of the chamomile family: *Anthemis tinctoria*, the Ox-Eye Chamomile, or Golden Marguerite; *Anthemis sanctijohannis*, the St. John Chamomile; *Anthemis cuparicaria*, the mat-forming chamomile; and *Anthemis arvensis*, the Corn Chamomile.

The action of the herb depends very much on the method of use, i.e. taken internally, neat or as a tissane, or applied externally. The form of the extract is of vital importance — whether it is an alcoholic extraction, an aqueous solution, a glycollic extraction, a pressed oil, the product of a steam distillation or indeed the dried or fresh herb itself.

The part of the plant used is also of great importance and the exact composition must be established (i.e. whether the flower, the petals, the seeds, the stem, the young leaves, the roots or the whole plant).

In addition, the species of the plant must be known and the geographical location, the time of collection, the method of storage and the reliability of the processor should be ascertained.

The choice of a good wine depends on the vine stock, the soil, the climate, the year and the country of origin. The same is true of a good extract since all the chemical elements and the proportions of their compositions will change according to the factors mentioned.

Finally, all oils should be analysed to ensure that they have not been adulterated by cheaper oils in order to obtain a better profit. The introduction of an Aromark grade (a seal of quality awarded by the Essential Oil Trade Association) is a welcome insurance to buyers of essential oils.

Twenty-four references were studied for their claims for chamomile, each chosen as representing the different spheres of interest, the nationality of author and a cross-section of the relevant industry.

A total of 108 different claims were found for chamomile, and a record for each claim was marked against the reference in which it occurred. Each claim was then scored for the number of times that it was reported.

Figure 1. English chamomile, *Anthemis nobile* L.

The top scores were as follows:

Antispasmodic	16
Antiseptic, wound-healing	15
Sedative-related	13
Anti-inflammatory	13

These were considered to have a high degree of confidence when made as a claim for chamomile.

The next group of claims had less conclusive scores and the claimed benefits were considered to be speculative:

Tonic	10
Soothing for conjunctivitis, sore eyes	10
Flatulence, carminative	10
Stomachic	9
Diarrhoea	8
Menstrual problems	8
Neuralgia	8
Insomnia, sleep-related problems	8

The remaining claims were considered to be inconclusive.

The claims were then compared to the original references using the simple calculation of the total number of claims made in that reference to the number of reported claims from the consensus groups above, expressed as a percentage. The purpose of the exercise was to assess the accuracy of the reference, since it was considered irresponsible to claim everything and anything, for all claims would hold equal weight in the eyes of the reader.

It is of interest to note that those books written by medical practitioners and herbalists tended to score more heavily than those written by the more fringe users of herbal extracts or those appealing to the more romantic reader.

The undeniable value of plant extracts will never be fully recognized until authors, suppliers, herbal users and marketing managers accurately and truthfully describe their natural materials, basing their information on substantiated data and exhaustive literature surveys.

Figure 2. German chamomile, *Matricaria chamomilla* L.

Table 1. Benefits

Condition	01	02	03	04	05	06	07	08	09
Medical-related									
Gallstone/Gall bladder disease									
Kidney stones/Kidney disease									
Haemorrhoids									
Liver disorders, Jaundice									
Urinary tract inflammation/ infection									
Earache							X	X	
Asthma (part treatment)									
Psoriasis									
Anorexia									
Eczema		X							
Dermatitis		X							
Anaemia									
Measles									
Shingles									
Urticaria									
Vaginitis									
Mastitis									
Cephalalgia									
Prostate problems									
Colitis									
Intestinal gas	X								
Flatulence									
Carminitive		X		X		X		X	X

attributed to chamomile

10	11	12	13	14	15	16	17	18	19	20	21	22	23
					X X								
X	X			X X			X			X		X	
X	X		X		X X X		X						
X	X X	X	X	X	X					X			
	X X X				X		X						X
X	X X X						X X X		X	X		X X X	X X

Table 1 —

Condition	01	02	03	04	05	06	07	08	09
Indigestion	X				X				
Stomachic						X	X		X
Dyspepsia								X	
Constipation									
Colic									
Gastritis									
Stomach disorders								X	
Peptic ulcers									
Stomach ulcers									
Cholagogue (bile producing)									
Aperitif									
Sharpen appetite							X		
Diarrhoea							X	X	
Sleep-related problems									
Insomnia	X			X					X
Nightmare					X		X	X	
Nerve-related problems									
Hysteria, nerves, relaxing									
sedative, anxiety, nervine, restlessness	X			X	X	X	X	X	
Hyperactivity									X
Depression									
Schizophrenia									
Frustration	X								
Irritability	X								
Epilepsy (part treatment)									
Hypnotic									

Continued

10	11	12	13	14	15	16	17	18	19	20	21	22	23
X	X X	X	X	X	X X		X	X	X X	X X		X	X X
X	X X		X				X					X	
X X	X X	X										X	
X X	X	X		X	X		X	X	X	X		X	X
X	X	X			X					X			
	X		X		X		X	X		X		X	
X X	X	X X			X X								
X	X	X X			X								

Table 1 —

Condition	01	02	03	04	05	06	07	08	09
Anger, oversensitiveness									
Pain-related									
General fatigue	X				X		X		X
Antispasmodic		X		X		X		X	X
Antiphlogistic		X							X
Analgesic									
Toothache					X				
Rheumatism, Lumbago Arthritis, Gout Headache, Migraine							X		X
Neuralgia							X	X	X
Menstrual problems, menopause, etc.							X		
Hangover							X		
Delirium tremens (D.T.'s)								X	
Skin-related									
General									X
Boils and abscesses		X						X	
Dermatitis		X							
Eczema		X							
Herpes		X							
Sores							X		X
Swellings							X	X	X
Skin ulcers						X			
Burns									

Continued

10	11	12	13	14	15	16	17	18	19	20	21	22	23
	X												
X	X												
	X	X	X	X	X		X	X	X	X		X	X
	X	X			X								X
	X	X											
X	X		X										
X	X	X											X
		X											
	X	X		X			X						
X	X	X	X		X								
X	X	X		X	X					X		X	
				X								X	
	X	X		X					X				
X	X												
		X								X			
		X											
		X											
							X	X		X		X	X
X	X	X											

Table 1 —

Condition	01	02	03	04	05	06	07	08	09
Broken veins									
Callouses									X
Rashes									X
Cicatrisant (scar-healing)									
Acne									
Antiseptic, speeds wound-healing		X		X	X		X		X
Anti-inflammatory				X			X	X	X
Dry, sensitive, hypersensitive, irritable skin									
Allergies									
Psoriasis									
Child-related									
Hyperactivity									X
Fretful babies	X				X				
Teething								X	
Colic									
Convulsions								X	
Ailments								X	
Disease									
Tantrums									
Measles									
Cosmetic uses									
Complexion wash	X				X				
Blond-hair rinse	X		X		X		X		

Continued

10	11	12	13	14	15	16	17	18	19	20	21	22	23
X													
X	X												
X	X	X		X			X	X	X	X		X	X
X	X	X		X			X		X	X		X	X
X													X
	X												
					X								
X		X			X								
X	X	X	X						X				
	X												
	X	X	X										
			X						X				
	X												
X													
					X								
						X							
						X				X		X	

Table 1 —

Condition	01	02	03	04	05	06	07	08	09
Oral-related									
Mouthwash for gingivitis, sore, inflamed gums				X					
Toothache					X				
Teething								X	
Aromatic									X
Ocular-related									
Conjunctivitis, sore, inflamed eyes, eyewash		X		X			X		
Various									
Insect repellent							X		
Bactericide									
Fungicide		X							
Vermifuge									
Helps to give up smoking									
Treats sweaty hands or feet									
Deodorant									X
Stimulant									X
Strengthens leucocytes		X							
Tonic						X	X	X	X
Anticatarrhal				X					
Diuretic									
Sudorific									
Febrifuge, fevers									X
Diaphoretic									X
Vertigo★									

★*Caused by chamomile in high doses*

Continued

10	11	12	13	14	15	16	17	18	19	20	21	22	23
	X			X						X		X	
X	X		X							X			
X	X	X	X						X				X
X	X	X		X				X		X			X
		X							X				X
		X											X
					X								
		X											
		X											
X	X	X	X									X	X
						X							
	X												
	X	X		X	X						X		
	X	X			X					X			
										X			
	X	X											

Table 1 —

Condition	01	02	03	04	05	06	07	08	09
Nausea★ Morning sickness, travel sickness, vomiting									

	01	02	03	04	05	06	07	08	09
TOTAL NUMBER OF CLAIMS	10	13	1	9	10	6	20	18	23
High-confidence claims, 13-16 responses - Group I	1	2	0	4	2	2	3	3	3
Percentage of claims	10	15	0	44	20	33	15	17	13
Speculative benefits, 8-10 responses - Group II	2	2	0	3	0	4	6	3	6
Group I and Group II	3	4	0	7	2	6	9	6	9
TOTAL PERCENTAGE OF CLAIMS - Groups I and II	30	31	0	78	20	100	45	33	39

★*Caused by chamomile in high doses*

In reference number order:

Ref. % Top
no. Claims

Ref. no.	% Top Claims	
01	30	Ceres. *Healing Power of Herbal Teas* (Thorsons, 1984)
02	31	Lautier, R. and Passebecq, A. *Aromatherapy — The Use of Plant Essence in Healing* (Thorsons, 1979)
03	0	Tisserand, M. *Aromatherapy for Women* (Thorsons, 1985)
04	78	Hoffmann, D. *Herb Users Guide* (Thorsons, 1987)
05	20	Leyel, C.F. *Herbal Delights* (Faber & Faber, 1987) and *Elixirs of Life* (Faber & Faber, 1987)
06	100	Mitton, M. *Herbal Remedies* "Skin Problems" and "Stress and Tension" (Foulsham, 1984)

Continued

10	11	12	13	14	15	16	17	18	19	20	21	22	23
	X						X			X		X	
				X								X	

33	50	39	13	14	26	2	15	8	12	21	1	19	18
2	4	3	2	3	2	0	4	3	4	4	0	4	3
6	8	8	15	21	8	0	27	53	33	19	0	21	17
6	7	6	3	3	4	0	2	2	2	5	0	3	4
8	11	9	5	6	6	0	6	5	6	9	0	7	7
24	22	23	38	43	23	0	40	63	50	43	0	37	39

07	45	Buchman, D.D. *Herbal Medicine — The Natural Way to Get Well and Stay Well* (Century Huchinson, 1987)
08	33	Grieve, Mrs M. *A Modern Herbal* (Savvas, 1984)
09	39	Lust, John. *The Herb Book* (Bantam, 1986)
10	24	Price, Shirley. *Practical Aromatherapy* (Thorsons, 1987)
11	22	Tisserand, Robert. *The Art of Aromatherapy* (C.W. Daniel, 1987)
12	23	Valnet, Dr J. *The Practice of Aromatherapy* (C.W. Daniel, 1986)
13	38	Wren, R.C. *Potter's New Cyclopædia of Botanical Drugs and Preparations* (C.W. Daniel, 1985)
14	43	Flück, Hans. *Medicinal Plants* (Foulsham, 1988)
15	23	Trattler, R. *Better Health Through Healing* (Thorsons, 1985)
16	0	Haxley, A. *Natural Beauty with Herbs* (Darton, Longman & Todd, 1977)
17	40	*British Herbal Pharmacœpia* (BHMA, 1983)

18	63	Mindell, E. *Vitamin Bible* (Warner Books, 1985)
19	50	Griffith, Dr D. Winter. *Vital Vitamin Fact File* (Thorsons, 1988)
20	43	Bunney, S., ed. *Illustrated Book of Herbs* (Octopus, 1984)
21	0	Wright, Michael. *Complete Handbook of Garden Plants* (Rainbird, 1984)
22	37	Stuart, Malcolm, ed. *Encyclopædia of Herbs and Herbalism* (Orbis, 1986)
23	39	Alban Muller data
24		Scheffer, M. *Bach Flower Therapy* (Thorsons, 1986)
25		Potterton, D., ed. *Culpepper's Colour Herbal,* (Foulsham, 1983)
26		*British Pharmacopeia Codex* (Pharmaceutical Press, 1973)

In claim order (see above for full details):

Ref. no.	% Top Claims	
06	100	Mitton, M. *Herbal Remedies*
04	78	Hoffmann, D. *Herb Users Guide*
18	63	Mindell, E. *Vitamin Bible*
19	50	Griffith, Dr H Winter. *Vital Vitamin Fact File*
07	45	Buchman, D. *Herbal Medicine*
14	43	Flück, Hans. *Medicinal Plants*
20	43	Bunney, S. *Illustrated Book of Herbs*
17	40	*British Herbal Pharmacœpia*, 1983
09	39	Lust, John. *The Herb Book*
23	39	Alban Muller data
13	38	*Potters New Cyclopædia of Botanical Drugs and Preparations*
22	37	Stuart, M. *Encyclopædia of Herbs and Herbalism*
08	33	Grieve, Mrs M. *A Modern Herbal*
02	31	Lautier, R. and Passebecq, A. *Aromatherapy — The Use of Plant Essence in Healing*
01	30	Ceres. *Healing Power of Herbal Teas*
10	24	Price, Shirley. *Practical Aromatherapy*
12	23	Valnet, Dr Jean. *The Practice of Aromatherapy*
15	23	Trattler, R. *Better Health Through Healing*
11	22	Tisserand, Robert. *The Art of Aromatherapy*
05	20	Leyel, C.F. *Herbal Delights* and *Elixirs of Life*
03	0	Tisserand, M. *Aromatherapy for Women*
16	0	Huxley, Alyson. *Natural Beauty with Herbs*
21	0	Wright, M. *Complete Book of Garden Plants*

Introduction to the Chemistry of Commercially Available Organic Natural Colours for the Cosmetic Formulator

Roy Nixon

Synopsis

The terms "natural", "nature-identical" and "natural origin" used in the colour industry are explained and the principal classes of natural colours — carotenes, chlorophylls and polyphenols — and the methods for extracting these classes of colours, are described. The properties of the colours, i.e. stability and solubility, are also explained, as are the pitfalls the cosmetic formulator may face when initially using natural colours.

The majority of colours used in the cosmetics industry at present are synthetic, but the use of and claims for natural colours are increasing. In the early 1980s the use of natural colours in food rapidly increased in response to consumer demand, and now in 1989 many natural colours which were thought to be too difficult (or unstable) to use are commonplace food ingredients. What, then, are natural colours and what is the fact or fiction related to their use?

The organic colours in the European Economic Community permitted lists for use in cosmetics may be broken down into four main sections, as follows:

Natural: colours extracted unchanged from nature by physical means, such as chlorophyll green from grasses.

Nature-identical: colours which occur in nature but can be chemically synthesized, e.g. the orange colour β-carotene.

Natural Origin: colours extracted from nature and chemically modified to form another species, such as the red cochineal carmine lake.

Synthetic: colours which have been synthesized and which do not occur in nature, for instance azo dyes have no natural counterpart.

Nature has surrounded us with an enormous range of natural colours covering the full spectrum and these can be divided into three main groups of interest: carotenoids, chlorophylls and polyphenols. Only the common commercially available types will be considered.

Carotenoids

Carotenoids are widespread in nature, being responsible for many of the yellow, orange and red colours found in vegetables, fruit, birds, fish and crustacea. Carotenoids are usually present in all green vegetation, their colour being masked by other colours, particularly chlorophyll.

It has been estimated[1,2] that nature produces around 100 to 200 million tonnes annually. Their function in the plant is thought to be as a photosynthetic action accessory and also a UV light absorber, protecting cells from damage; thus, they act as *nature's own sunscreens.*

The carotenoids are unsaturated hydrocarbons belonging to the terpene group with a long-chain conjugated double bond system. There are at least three hundred carotenoids which can be split into two groups: the xanthophylls and the carotenes.

The carotenes are hydrocarbons; the closely related xanthophylls also contain oxygen. Examples are shown in figure 1. The apo-carotenoids are xanthophylls from which fragments have been lost from either one or both ends of the chain.

Sources

The most commonly available natural carotenoid extracts are tabulated below (see table 1), together with the main source of each.

Extraction

The selected part of the plant which contains the colour, i.e. the outer pericarp of the paprika seed pod, the root of the carrot, etc., is isolated and cleaned.

Table 1. Main natural carotenoid extracts and their sources

Main extracted carotenoids	Sources	Colour
Crocin } Crocetin }	saffron (crocus stigma); gardenia fruit	yellow
Bixin } Norbixin }	*Bixa orellana* L. (annatto tree seeds)	yellow/orange
Capsanthin } Capsorubin }	paprika pericarp (*Capsicum annum*)	orange/red
Lycopene	tomato	red
Lutein	marigold (*Tagetes erecta* L.)	yellow
Carotene } β-Carotene }	algae (*Dunaliella salina*); carrots	yellow/orange

It is then macerated and carefully dried. Solvent extraction is used to isolate the carotenoid pigments which, being hydrocarbons, are soluble. The clarified solvent is then removed under reduced pressure to leave a resinous material containing the colour. This process is known as an oleoresin process and the resulting coloured product is known as an oleoresin, e.g. oleoresin paprika, oleoresin carrot, etc.

Nature-identical carotenoids are prepared from pure chemicals following defined synthetic pathways and are exact replicas of the natural compound. Although other carotenoids have been produced, only β-carotene, β-apo-8′ carotenal and canthaxanthin are regularly commercially available.

A water-soluble carotenoid of *natural origin* can be prepared from bixin (the diapo-carotenoid extracted from *Bixa orellana* tree seeds). Bixin is a monomethyl ester of a dicarboxylic acid and

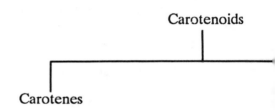

α-carotene

β-carotene

Figure 1.

Xanthophylls

lutein

capsanthin

capsorubin

if it is reacted with potassium hydroxide a water-soluble di-potassium salt of norbixin is formed (see fig. 2).

Properties
Colour
The long chain of conjugated double bonds present in these compounds is responsible for the colour and, in general, the longer the conjugated chain (chromophore) the redder the appearance (see fig. 3 for examples).

Solubility
Carotenoids, being hydrocarbons, are generally insoluble in aqueous media, the notable exceptions being annatto and saffron extracts which have carboxylic acid groups in their structure and are soluble in alkaline conditions only.

They are normally supplied as suspensions of micronized crystals in vegetable oils for further dilution and dissolution by the customer.

The colour industry has produced a variety of emulsified and colloidal suspension forms which give water dispersibility by using a variety of protective colloids and surface-active agents. The soluble annatto norbixin salts are insoluble in acid conditions.

Heat
Most carotenoids occur in the all-*trans* form, which on prolonged strong heating changes to the *cis*-isomer form with a consequent yellower shade change and reduction in intensity. They will not show any significant loss of colour if boiled for short periods, however.

Light exposure will also promote *trans/cis* isomerization and consequent loss of intensity. Also, in the presence of oxygen, light will catalyse the oxidation of the conjugated double-bond chromophore system. Partially oxidized carotenoids, such as canthaxanthin and bixin, are surprisingly more stable to air.

The protection of carotenoids against oxidation is, therefore, most important. This may be achieved by using sealed containers, minimal headspace, vacuum packing and opaque packaging. Also, the oil-dispersed forms of the colours are more stable than

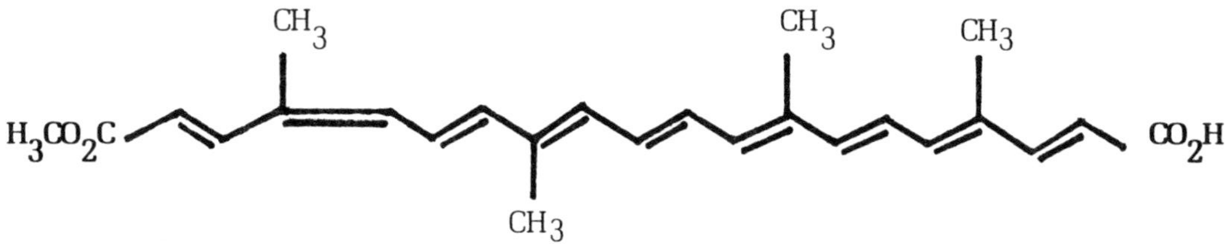

Bixin

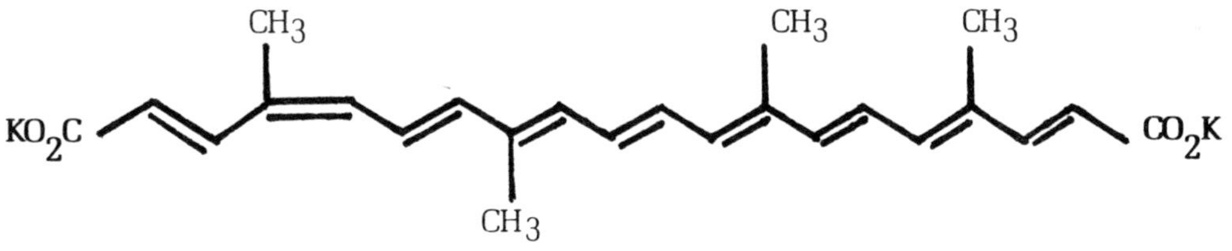

Norbixin di-potassium salt

Figure 2. Carotenoids: bixin and di-potassium norbixin

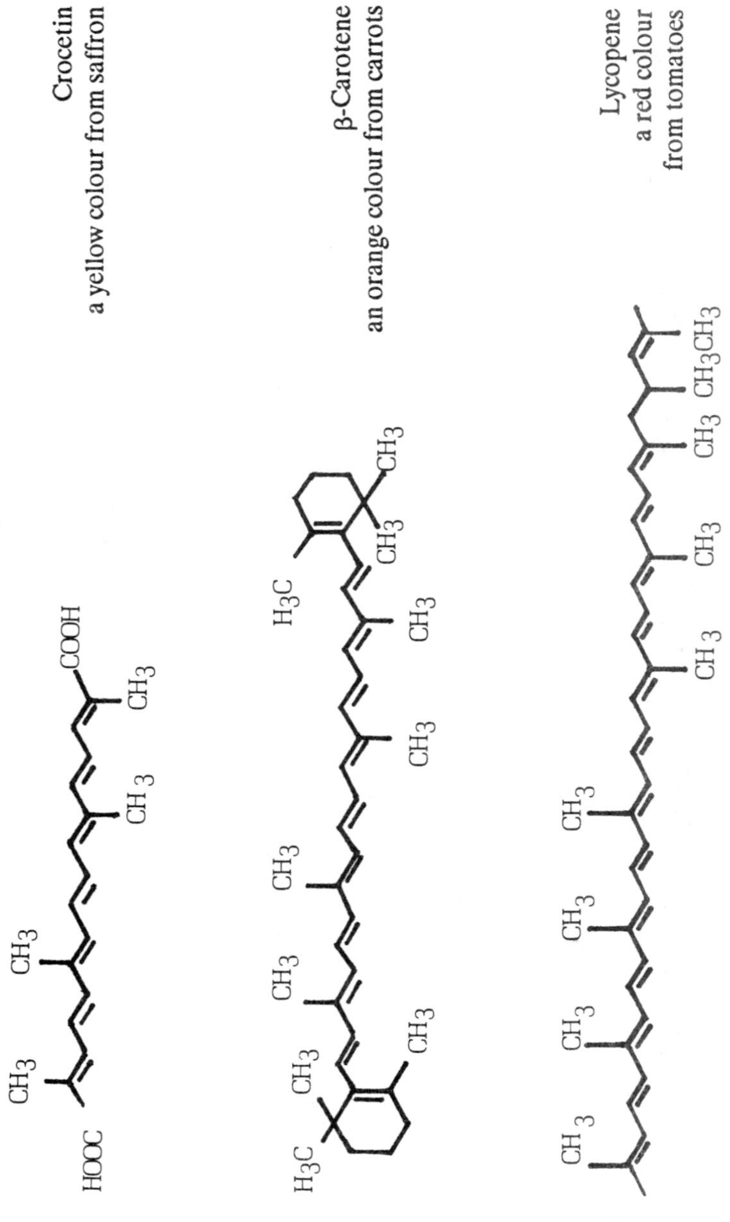

Crocetin
a yellow colour from saffron

β-Carotene
an orange colour from carrots

Lycopene
a red colour
from tomatoes

Figure 3. Carotenoids: crocetin, β-carotene and lycopene

aqueous dispersions. The addition of antioxidants is also recommended: BHA, BHT or α-tocopherol for non-aqueous systems, and ascorbic acid or sodium pyrophosphate for aqueous formulations.

One pitfall the formulator may face with skin-care preparations in particular is to claim that a product including oleoresin carrot or β-carotene contains vitamin A. β-Carotene has, in common with several other carotenoids, pro-vitamin A activity. This means that the carotenoid molecule is capable of forming vitamin A (retinol) in the digestive system but it is not in itself vitamin A (see fig. 4). The carotenoid must contain at least one β-ionone ring to be a vitamin A precursor.[3]

Chlorophylls or Porphyrin Pigments

Chlorophylls are nature's most abundant pigment.[2] It is estimated that around 1.2 billion tonnes per year are produced by plants and algaes over the entire globe. They are present in all photosynthetic cells and necessary for the conversion of carbon dioxide and water into carbohydrates using the light energy from the sun. Chemically, chlorophylls are derivatives of porphyrin, which are cyclic compounds containing four methine-linked pyrrole rings, plus a fifth ring formed at position 6.

Magnesium Chlorophyll

It can be noted from the structure that it is a co-ordination complex with magnesium as the central atom and also that there is a long-chain isoprenoid alcohol attached to the C7 propionyl substitute. This is known as the phytol side chain and enhances the molecule's oil solubility.

There are several types of chlorophylls based upon the chemical structure above, but in green plants from which chlorophyll is extracted for colour use, types *a* and *b* are present. For type *a*, the substitute R is $-CH_3$; for type *b*, a more yellowish green, R is -CHO. The ratio varies from plant to plant but type *a* predominates at 2 to 3 parts to 1 of *b*.

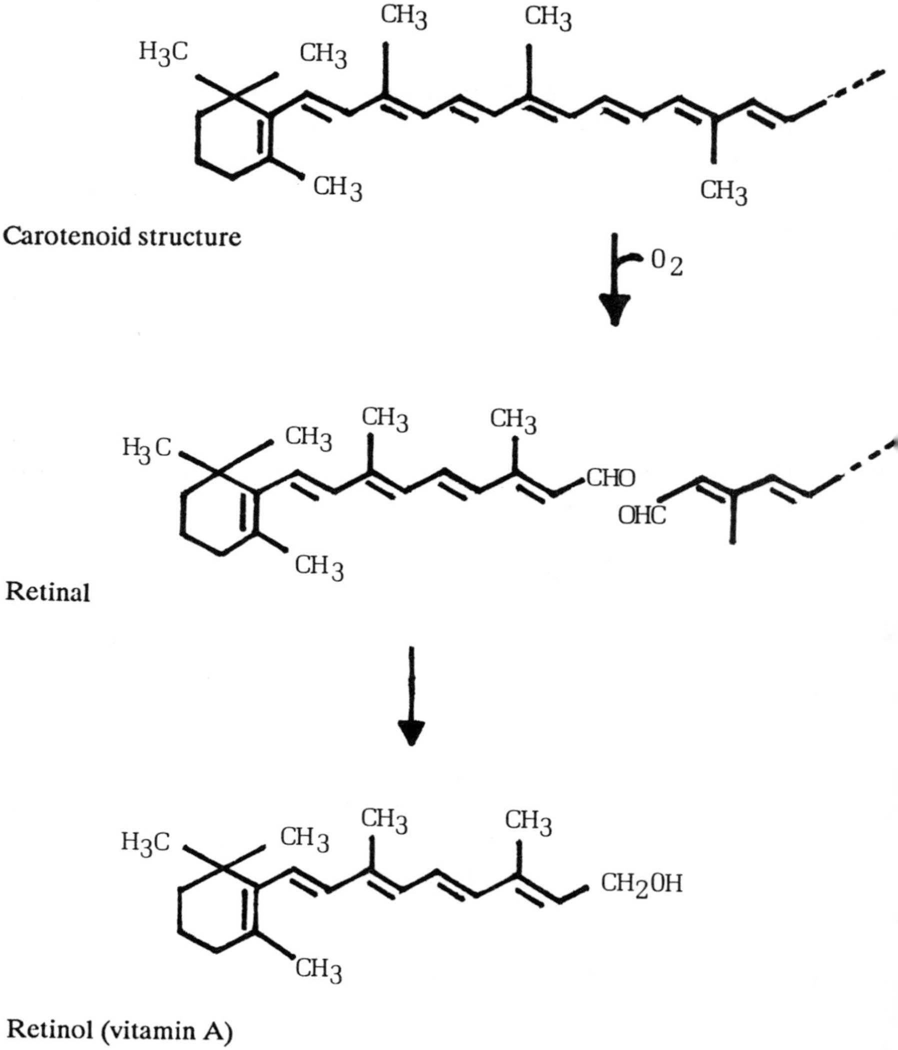

Carotenoid structure

Retinal

Retinol (vitamin A)

Figure 4. Formation of retinol and retinal

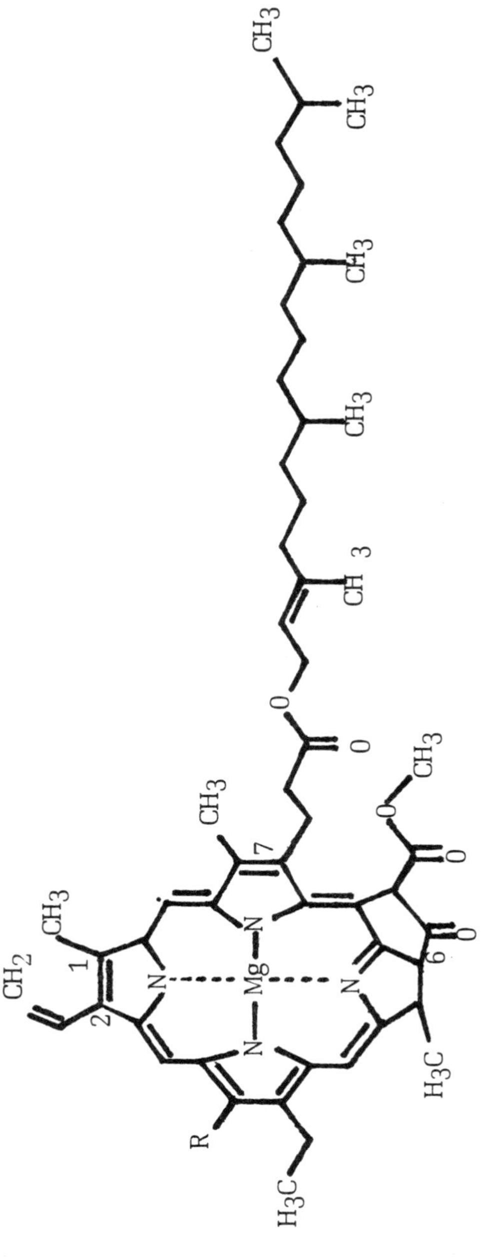

R = CH$_3$ for chlorophyll *a* and R = CHO for chlorophyll *b*

Figure 5. Magnesium chlorophyll

Extraction

Chlorophylls are extracted from lucerne grasses. These grasses are harvested, macerated, dried and then selectively solvent-extracted to remove chlorophyll from the phospholipids, fats and carotenoids (mainly xanthophylls) present. After evaporation of the solvent, the oil-soluble colour is standardized with vegetable oil. The conditions of the extraction have to be strictly controlled to avoid loss of magnesium from the molecule with the formation of pheophytins, which have a brownish colour.

Natural origin colours are produced from chlorophyll in two ways:

1. By reaction of the chlorophyll extract with a copper salt in mildly acidic conditions, the magnesium central atom is easily replaced by copper to produce copper chlorophyll. Copper chlorophyll is a more stable and brighter, bluer green than the more khaki-green natural chlorophyll extract.

2. Hydrolysis of the oil-soluble chlorophyll in mildly alkaline conditions results in the removal of the phytyl ester from the C_7 propionate group. The resulting sodium salt is known as a chlorophyllin and is water-soluble. A water-soluble copper chlorophyllin is also available (see fig. 6).

Properties

Colour

Magnesium chlorophylls and chlorophyllins are a dull khaki green. Copper chlorophylls and chlorophyllins are a brighter, bluer green.

Solubility

Chlorophylls and copper chlorophylls are oil-soluble only. The chlorophyllins are water-soluble only in alkaline conditions. In acid conditions the free acid will precipitate, so acid-stable forms of the chlorophyllins have been developed.

Heat

Magnesium chlorophylis, if heated, especially in mildly acidic conditions, will lose further Mg^{2+} ions and turn a dirty brown colour. In alkaline conditions, as the housewife knows, the

Magnesium
chlorophyllin

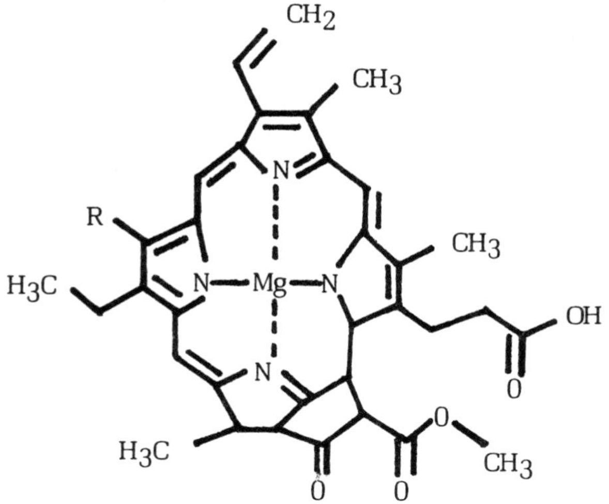

R = CH$_3$ or CHO
according to
chlorophyll type

Copper
chlorophyllin

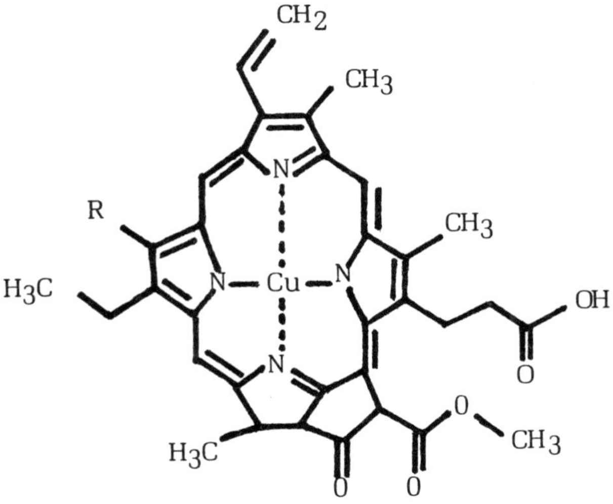

Figure 6. Chlorophyllins

molecule is stabilized, and cabbage, for example, if cooked with a pinch of sodium bicarbonate, stays green. Copper chlorophylls and chlorophyllins show good stability to heat.

Light
Stability to light is good; the copper complexes are very good.

Metals
Stability of the uncoppered chlorophylls and chlorophyllins to metals is poor. The central atom is easily replaced by other metal ions, particularly iron, which can give a reddening effect with time. The colour produced is related to haem, the red colour present in red blood cells. The copper complexes are stable to other metal ions.

Odour
Chlorophyll does have a grassy odour if used at high concentrations, but at normal levels of use this is not noticeable.

Polyphenols

The polyphenolic colours that occur in nature are a very varied group that have some water solubility and occur as glycosides.

A glycoside is a sugar-component molecule linked to a non-sugar component (in our case the colour chromophore) by a glycosidic link. The non-sugar component molecule is called an aglycone.

The polyphenolic colours divide roughly into three groups: quinonoids, betalaines and anthocyanins.

Quinonoids
These red and brown pigments are widely distributed in nature and occur in plant roots, wood, wood bark and certain insects.

Cochineal
Cochineal is a red quinonoid colour extracted from the dried bodies of the *Coccus cacti* female beetle. This beetle is a parasite which lives on cactus plants in hot dry areas of the world, such as Peru, Mexico, Guatemala and the Canary Islands.

Extraction

The dried bodies of the insect are milled to a powder and then extracted with aqueous alcoholic solution. After clarification the solvent is evaporated to leave cochineal. Precipitation, rewetting with solvent and evaporation may be done several times to produce a purer product. The pigment present in cochineal is a polyhydroxy-anthraquinone glycoside, called carminic acid, which has the structure:

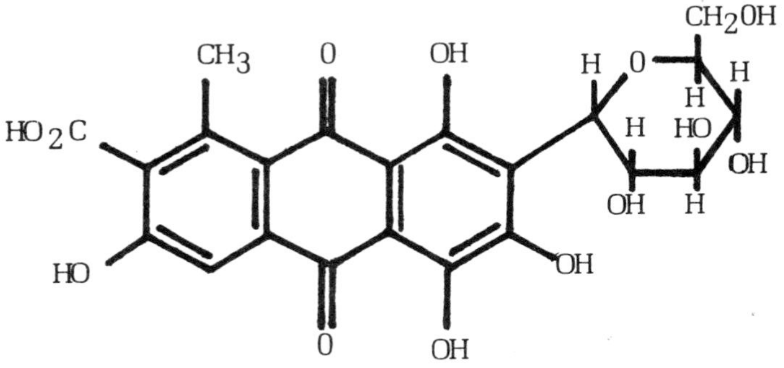

Figure 7. Carminic acid

The *natural origin* lake colour cochineal carmine lake is prepared from the aqueous extract by adding an aluminium salt and precipitating out the aluminium/carminic acid complex with alcohol. The cochineal carmine lake is thought to have the structure shown in figure 8.

Properties

Colour

In slightly alkaline solution cochineal is a blue-red resembling amaranth. According to customer requirements, cochineal carmine lake can be supplied in yellowish red to a blue-red shade.

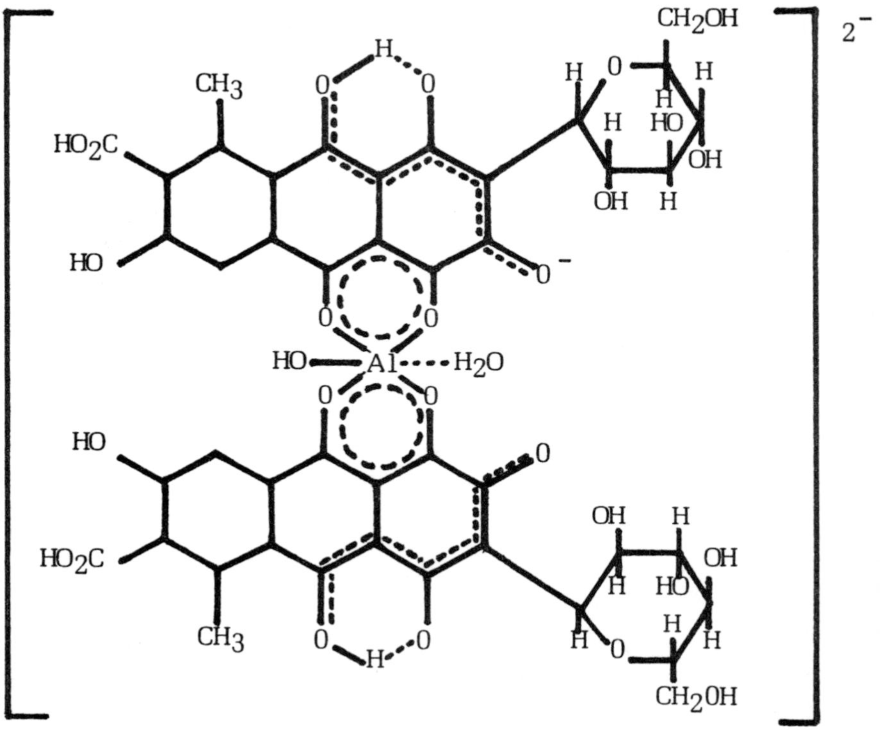

Figure 8. Cochineal carmine lake

Solubility

Cochineal is soluble in aqueous solutions above pH 3. Below this pH the free acid will precipitate as a red pigment. Cochineal carmine lake is not soluble in alcohol, glycerine, water or dilute acidic media, but solubilization will occur in alkaline solutions. In fact, carmine is available as solutions in ammonia and aqueous alkalies.

Heat

Stability of cochineal and cochineal carmine lake to heat is very good.

Light

Again, stability of both forms to light is very good.

Cost

The cost of cochineal products is high, probably due to the fact that it takes about 140,000 female insects to produce 1 kg of crude pigment. However, for a natural colour it is very strong and stable.

Odour

None.

Betalaines

Betalaines are a group of water-soluble pigments found in red beets (*Beta vulgaris* L.). This class of colour splits into two, the betacyanins and the betaxanthins. The major red pigment in red beet is the glycoside betanin, a betacyanin, but there are also present two minor yellow colours, vulgaxanthin I and II (see fig. 9).

It can be seen that the structures are similar up to the nitrogen atom, but that the betanin structure has increased conjugation, which causes it to undergo a bathochromic shift and appear red.

Extraction

Colour is present in the fresh root at about 0.1 per cent of its weight. The washed root is crushed and the juice expressed is

Betanin

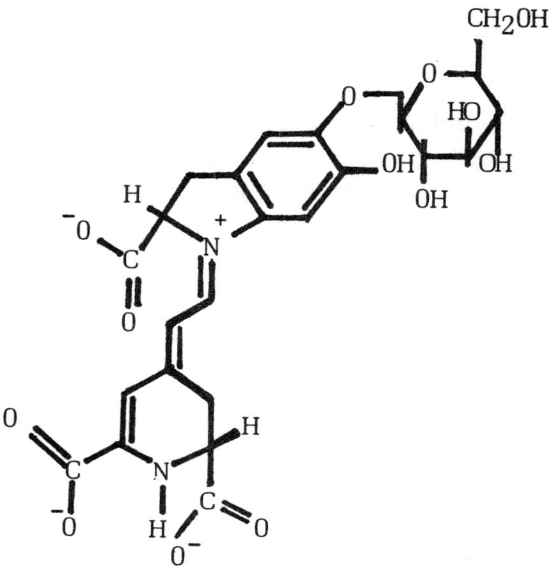

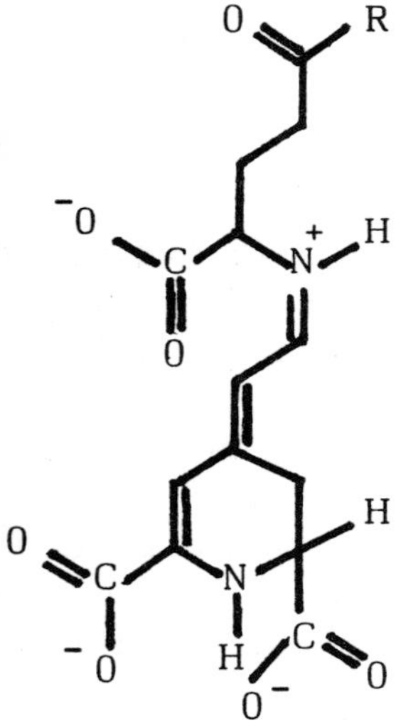

Vulgaxanthin I, R = NH_2
Vulgaxanthin II, R = OH

Figure 9. Betalaines

acidified with citric acid. The product is then sold as a liquid concentrate or spray-dried to a powder on a dextrin carrier.

Properties
Colour
A deep bluish red in aqueous solution.

Solubility
Beetroot extract is soluble in aqueous solutions and most stable between pH 4 and 5. Only very slight colour changes will occur between pH 2 and 9 but above 10 yellowing will occur. It is insoluble in oils and fats.

Light
Light stability is best at a pH between 4 and 5. Outside this range stability is poor. Oxygen elimination also helps prevent deterioration.

Odour
A slight beetroot odour is perceptible from both the liquid and powder forms. Low-odour grades are available.

Metals
Colour stability is reduced in the presence of metal ions, particularly iron and copper. The addition of the sequestering agent EDTA improves stability.

Moisture
The powder form is hygroscopic and will tend to cake if exposed to the air.

Anthocyanins or Flavonoids
The anthocyanins are the largest group of water-soluble natural colours and they are responsible for the attractive colours of fruits and flowers as well as leaves of some plants. The colours range from yellowish reds, through purples to deep blues. They all occur as glycosides and some 160 different anthocyanins are known to exist. Space does not permit us to go into the detailed chemistry

of the anthocyanins but if we look at the structure[4] of the six major anthocyanin aglycones (called anthocyanidins), we can explain their properties.

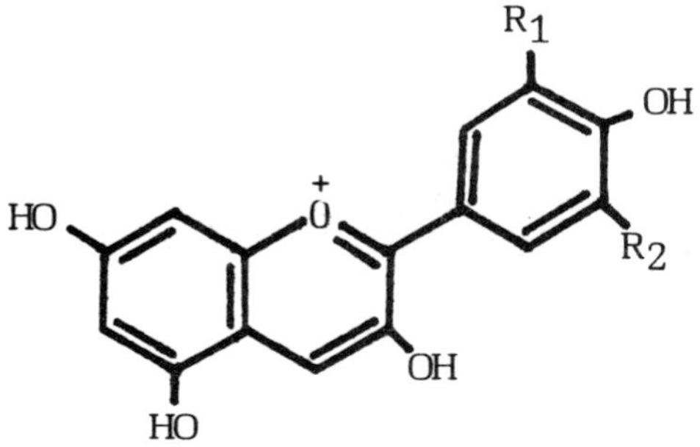

	R_1	R_2	λ_{max} nm	Red
Pelargonidin	H	H	520	
Cyanidin	OH	H	535	
Peonidin	OCH_3	H	532	
Delphinidin	OH	OH	546	
Petunidin	OCH_3	OH	543	
Malvidin	OCH_3	OCH_3	542	
				Blue

Figure 10. Major anthocyanidins

An increase in hydroxylation or methoxylation of the aglycone results in an increased blueness. The shade of anthocyanin colourants is therefore source-dependent as most fruit plants contain mixtures of anthocyanins.

Extraction
The most common commercially available anthocyanin colour is extracted from the red European grape *Vitis vinifera*, which contains a cyanidin, peonidin, delphinidin, petunidin and malvidin mixture.

Grape pomace, a by-product of the red wine industry, is extracted with alcohol containing citric acid and a small quantity of sulphur dioxide. The liquid product may then be standardized for direct sale or converted into a powder by spray drying.

Properties

Colour and solubility

Anthocyanins are freely soluble in water and alcohol but they are insoluble in fats and oils. One factor to keep in mind is that the colour of anthocyanins in aqueous solutions is very dependent on pH. They exhibit their typically red hue at very low pH's of 1 to 3, gradually weakening and becoming purple as neutrality is approached and finally blue in the range 7 to 10. The mechanism of what is happening is not clearly understood for mixtures of anthocyanins, but studies[5] have been made on purified single species and the explanation seems to be transitions in the chromophore's structure as outlined in figure 11.

1. At pH 1 and below, anthocyanins exist as red flavium salts.
2. As the pH is raised, a proton is lost, a water molecule is gained and a colourless carbinol pseudo-base is formed — hence the weakening.
3. Above pH 3, anthocyanins in the flavium cation form begin to lose protons, forming the quinonoidal base which is purple.
4. Also above pH 3, some of the carbinol pseudo-base is gradually converted into a colourless chalcone form.
5. At higher pH's around 7 (when further protons are lost), the blue ionized quinonoidal base is formed.

These variations in chemistry and shade emphasize that strict pH control is essential when using this colour. It is also interesting to note the loss of conjugation in the structures of the two colourless forms of the chalcone and carbinol pseudo-base.

Heat

Prolonged heating in the presence of oxygen will cause some colour loss.

Flavylium cation (AH$^+$)

Quininoidal base (A)

Ionized quininoidal base (A$^-$)

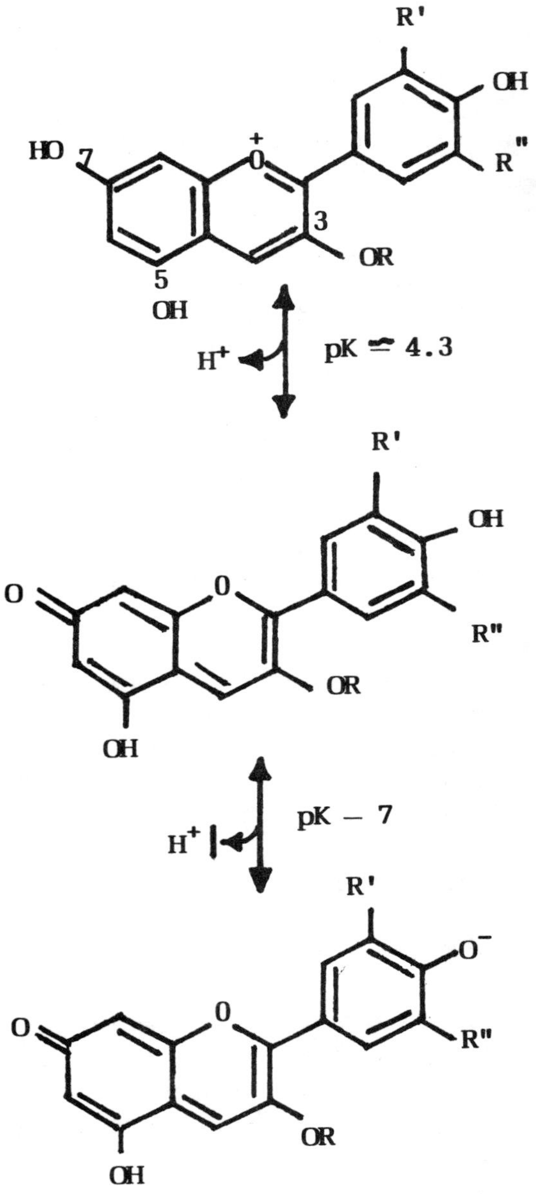

Figure 11. Transitions in

Carbinol
pseudo-base (B)

$pK \rightleftharpoons 3$

Chalcone (C)

anthocyanin structure

Light
Anthocyanin has good light stability.

Metals
Anthocyanins are excellent chelators of metals, which cause colour changes even at low levels. The food industry has cured the problem by using lacquer-lined cans, and the cosmetic formulator contemplating using this colour must guard against metal contamination.

Moisture
The powder is very hygroscopic and will become sticky if left open to the atmosphere.

Summary
It is hoped that this paper shows that a natural pigment can offer an alternative to some of the synthetic colours available and that for successful application a basic understanding of the chemical and physical properties of the natural colour under evaluation is required. Although only a few of the common commercially available natural colours have been dealt with in this paper, many others are available and the advice to the formulator is to study the chemistry before contemplating experiments.

The food industry, in response to consumer demand about eight years ago, has now managed to use natural colours extensively as replacements for synthetic colours, whereas previously it was assumed that natural colours were unusable. Will the future see more natural colours in cosmetics?

References
1. Walford, John, ed. *Developments in Food Colours — 1*. London: Applied Science Publishers.
2. Hendry, George. "Where Does All the Green Go". *New Scientist* (November 5, 1988).
3. Counsel, J.N., ed. *Natural Colours for Food and Other Uses*. London: Applied Science Publishers.
4. Engel, C. *Natural Colours, Their Stability and Application in Food.* Leatherhead Food Research Association.

5. Coultate, T.P. *Food — The Chemistry of its Components*. 2nd ed. London: Royal Society of Chemistry.

Usnic Acid — A Natural Deodorant?

Malcolm James and David Mitchell

Introduction

Bacteria occupy a very unique position in the environment. Their adaptability has meant that species can survive conditions that we would find intolerable, but they are so small that their presence usually remains undetected unless there is some macro effect, for example, food turning mouldy, people becoming ill, offensive odours, etc.

One "offensive" odour we are all familiar with is that produced by the microflora present on our skins. This population of bacteria, fungi and yeasts utilize our excretions as sources of food. We call these excretions sweat or perspiration.

The human body has two types of sweat glands, eccrine and apocrine. Eccrine glands respond to changes in the environment and to changes within the body due to physical activity. They function as part of the human body's temperature regulation system. Eccrine glands secrete mostly water and are not usually the cause of body odours. Apocrine glands, on the other hand, secrete lipids and proteins, and are linked more to the body's emotional responses. They are active when we are nervous or excited, and are predominantly found under the armpits. It is the breakdown products of these fatty materials and proteins which give rise to the offensive odours.

There are three ways of reducing these odours:

a) by masking the smells with perfumes;
b) by controlling the amount of sweat produced by the body, or
c) by reducing the population of micro-organisms present on the skin.

The products marketed to control body odour come in two forms:

> a) antiperspirants — typically a mixture of an aluminium or a zirconium salt to control perspiration, a bactericide (or a bacteriostat) and a fragrance; or
> b) deodorants — perfume oils, and possibly materials to control bacterial growth.

Deodorants have been traditionally a neglected product area, only recently enjoying success in the United Kingdom with the introduction of the deo-cologne and an emphasis on the marketing of fragrances and their effects on close personal contacts.

The bacterial population on our skins contains many different species of organism, mainly gram-positive bacteria such as the group *Staphylococci*, and particularly *Micrococcus albus*. The organisms are not easily removed by washing with soap and water since they live not only on the skin surface, but also in the hair follicles and sebaceous glands. To eliminate these hidden bugs requires an antibacterial agent. In the past some work has been done to evaluate the possibility of using antibiotics for this purpose, although it is now recognized that there are two reasons why they should not be used:

> a) exposure of bacteria to antibiotics builds up resistant strains; and
> b) skin disinfection can lead to problems with pathogenic microbes.

The materials used in deodorants to control bacterial growth are generally bacteriostats rather than bactericides, and, in use, levels are kept to a minimum to avoid the complications outlined above. Deodorants are intended to perform a cosmetic rather than a pharmaceutical function.

It should be noted that some of the materials used to prepare fragrances are capable of acting as bacteriostats, for example:

Phenylethyl alcohol	rose
Benzyl alcohol	rose
Thymol	thyme
Eugenol	carnation
Isoeugenol	carnation
Citral	lemon
Cinnamic aldehyde	cinnamon

However, the active ingredient in deodorants is more likely to be a synthetic chemical of some kind, for example Irgasan DP 300 (triclosan) from Ciba-Geigy or 3,4,4-Trichlorocarbanilide (triclocarban). A possible natural alternative is provided by a plant called Alpine lichen.

Description

Alpine lichen is usually to be found hanging from the branches of trees on the slopes of the Alps. In appearance it is a mass of fine threads. The active ingredient of Alpine lichen, and of other related species, is called usnic acid.

Usnic acid ($C_{18}H_{16}O_7$) was first extracted in 1843 by Rochleder and Heldt. It is a dibenzofuran and its formula is illustrated below along with its chemical name:

2,6-Diacetyl-7,9-dihydroxy-8,9b-dimethyl-1,3(2H,9bH)-dibenzofuran-1-one.

Usnic acid is an optically active, yellow coloured material, which crystallizes in a characteristic needle form, with a melting point of around 200°C. The isomer derived from *Usnea barbata* is the dextrorotatory form, but the laevorotatory form is also found in nature, as well as the racemic mixture.

The solubility characteristics of usnic acid are listed in the following table:

Table 1. Solubility of usnic acid

Material	% w/w
Water	<0.01
Acetone	0.77
Chloroform	4.60
Ethanol	2.12
Isopropyl alcohol	0.28
Triglyceride	0.30

As can be seen, the pure acid is not very soluble at all, but this disadvantage can be overcome by using its salts. The sodium salt is more soluble in water at pH 9, and completely soluble in mixtures of ethanol and propylene glycol. Further improvements can be achieved by using amine salts.

Activity
The following table lists the minimum inhibitory concentration against a range of organisms:

Table 2. Minimum inhibitory concentration of usnic acid

Species	ATCC No.	Conc. mcg/l
Gram (+)		
Bacillus subtillus	6633	0.9-2.7
Bacillus cereus	9634	0.9-2.7
Bacillus coagulans	12245	0.9-2.7
Staphylococcus aureus	6538p	8.2-25
Sarcina lutea	9341	2.7-8.2
Streptococcus faecalis	8043	2.7-8.2
Gram (-)		
Pseudomonas aeruginosa	25619	666
Escherichia coli	9637	666
Fungi and yeasts		
Candida albicans	10231	25-74
Aspergillus niger	1015	2.7-8.2
Aspergillus flavus	9643	0.9-2.7
Penicillium notatum	9178	0.3-0.9

Usnic acid and its salts are not active against gram-negative bacteria, but they are very active against gram-positive bacteria, fungi and yeasts. The literature indicates that the mechanism for its activity is the inhibition of oxidative phosphorylation, thus limiting mitosis and cell respiration.

Work has been done to compare usnic acid with imidazolidinyl urea. Since the latter is probably very familiar to the cosmetic industry, we include the results here. The table shows diameters of zones of inhibition in millimetres. The larger the diameter, the more effective the material is at preventing growth.

Table 3. Comparison of inhibitory zones of usnic acid and imidazolidinyl urea

| Species | Inhibitory Zone Diameters /mm | |
	Usnic Acid	Imidazolidinyl Urea
Bacillus subtillus	43	39
Staphyl. epidermis	39	36
Sarcina lutea	39	25
Pseudomonas aeruginosa	0	18
Aspergillus flavus	19	0

Test concentrations:
Usnic Acid: 0.04% of the amine salt
Imidazolidinyl Urea: 2.00%

As the table shows, usnic acid is at least comparable to imidazolidinyl urea in activity. There is a possibility that a combination of the two might serve as a product preservative.

Stability
The amine salt of usnic acid is susceptible to discolouration — darkening — with time, an effect which should be taken into account when formulating products. The following materials and situations should be avoided:

> Titanium dioxide
> Cationic materials
> Acid pH
> Copper or other metal salts
> Alkali metal salts

The activity of usnic acid and its salts is unaffected by the presence of nonionic and anionic ingredients, as illustrated by the following work. Formulations as listed below were prepared, and these were then compared with a standard usnic acid solution.

a) *Anionic surfactant*

   5% solution of Usnic Acid in Dipropylene Glycol   4%
   Sodium Laureth Sulfate (28%)   40%
   Water   56%

b) *Anionic emulsion*

   5% solution of Usnic Acid in Dipropylene Glycol   4%
   Triethanolamine Stearate   40%
   Water   56%

c) *Nonionic solubilizer*

   5% solution of Usnic Acid in Dipropylene Glycol   4%
   PEG-40 Hydrogenated Castor Oil   40%
   Water   56%

Table 4. Activity of usnic acid in the presence of anionic and nonionic ingredients

Bacillus subtilis

| | Inhibitory Zone Diameters/ mm | |
	Reference	37°C for 24 days
Anionic surfactant	21	21
Anionic emulsion	23	23
Nonionic solubilizer	19	18

Staphylococcus epidermis

| | Inhibitory Zone Diameters/ mm | |
	Reference	37°C for 24 days
Anionic surfactant	21	21
Anionic emulsion	23	23
Nonionic solubilizer	17	14

Sarcina lutea

| | Inhibitory Zone Diameters/ mm | |
	Reference	37°C for 24 days
Anionic surfactant	22	22
Anionic emulsion	24	23
Nonionic solubilizer	13	12

Toxicology

A 5 per cent solution of the amine salt of usnic acid in 1,2 dipropylene glycol was found to be slightly irritating to eyes, but not a skin irritant, nor was it toxic. The LD50 for this solution was in excess of 10 g/kg. Other researchers have found the following LD50 values for the pure acid:

Mikoshiba (1936)

mice 25 mg/kg — intravenously
rats 30 mg/kg — intravenously
dogs 40 mg/kg — intravenously

Schimmer (1974)

mice 290 mg/kg

There has been some evidence recently to suggest that usnic acid could be linked to oakmoss allergy.[1] Further tests indicate that usnic acid is a very weak sensitizer.[2] Recently the Swiss Health Federation approved the use of usnic acid (and its salts) at levels up to 0.5 per cent in skin cosmetics. No legislation appears to exist within the European Economic Community (particularly Germany) to limit its use in cosmetics.

Could usnic acid be used in a deodorant? Let us review what we know about this material:

1) Activity
Usnic acid is active against gram-positive but not gram-negative bacteria. Since gram-positive bacteria are prevalent on the skin this should be acceptable.

2) Toxicology
Some evidence to indicate sensitivity, possibly an area where further work is needed. At present the literature points to usnic acid being a weak sensitizer.

3) Use as a drug
Usnic acid could be used as an antibiotic to treat tetanus, meningitis and diphtheria. However, the authors believe that usnic acid is not being used for this purpose at the present time, so cosmetic use remains unrestricted.

4) Compatibility
There is quite an array of materials (and conditions) with which usnic acid is not compatible. Careful work would be needed to formulate an acceptable formulation.

5) Stability
Solutions of usnic acid salts are not very stable. Work is progressing to overcome this difficulty. Formulators should liaise with marketing and discuss packaging, etc.

Having considered the above points, the authors believe that usnic acid is probably worthy of consideration for the proposed application. The following formulations are given as examples:

Deodorant spray

	%w/w
Deo-Usnate*	0.9
Perfume	0.9
Propylene Glycol	3.0
Herb. Extract Balm Mint (IPA)	2.0
Ethanol 96% v/v	26.5
Propellant	66.7

Foot spray

	%w/w
Deo-Usnate*	1.0
Herb. Extract Witch Hazel (IPA)	2.0
Perfume	1.0
Menthol	0.2
Propylene Glycol	5.0
Ethanol	80.8
Demineralized Water	10.0

Deodorant stick

	%w/w
Deo-Usnate*	1.50
Herb. Complex Alpine Herbs	2.00
Perfume	1.70
Sodium Hydroxide	1.33
Stearic Acid	8.5
Propylene Glycol	68.70
Demineralized Water	16.27

*Deo-Usnate is a lichen extract in 1,2 Dipropylene Glycol containing 5% usnic acid.

References

1. *Contact Dermatitis,* Dec. 1988
2. Hausen, Bjoern M. *Allergiepflanzen, Pflanzeneallergene.* Ecomed Landsberg. Munich: 1988.

Bibliography and General References

Fearnley and Cox. "A New Microbiological Approach to the Assessment of Underarm Deodorants." S.C.S. Symposium, "Microbiology and Toxicology", 1981

Wilkinson, J.B. and Moore, R.J., eds. *Harry's Cosmeticology,* 7th edition. London: Longman, 1982.

Rothwell, P.J. "Deodorant Activity and Evaluation." S.C.S. Symposium, "Sensory Evaluation of Product Performance", 1980.

Baxter and Reed. "Evaluation of Deodorants." S.C.S. Symposium, "Substantiation of Cosmetic Benefits", 1982.

Hart and Hutchinson. "The Effect of Some Toiletry-Grade Surfactants on the Skin's Microflora." I.F.S.C.C. Congress, Barcelona, 1986.

The Merck Index, 10th edition. Edited by Martha Windholz. Rahway, NJ: Merck & Co., 1983.

Culbertson. *Chemical and Botanical Guide to Lichen Products.* University of North Carolina Press, 1969.

Asahina and Shibatas. *Chemistry of Lichen Substances.* Amsterdam: Asher & Co., 1971.

Fontana. "Usnic Acid." Italian Society of Cosmetic Scientists lecture, 1974.

Nowak, G.A. *Cosmetic Preparations,* Vol. 1, 3rd edition. Augsburg, Germany: H. Ziolkowsky, 1985.

Cosmetochem AG, "Deo-Usnate, A Bactericide." *Cosmetics & Toiletries,* Vol. 98, No. 6 (June 1983).

Gongalo, Cabral and Gongalo. "Contact Sensitivity to Oak Moss." *Contact Dermatitis,* Dec. 1988.

Feger. "Untersuchung zur konservierenden wirkung von stoffen naturlichen ursprungs." Kreuzncher Symposium "Konservierung Kosmetischer Mittel", 1986.

Index